浙江省级一流本科课程配套教材

线上线下混合式计算机图形学基础实验教程

潘万彬　王毅刚　曹伟娟　王若兰
洪雨洁　黄婷婷　徐纯芝　董琛琛　编　著

西安电子科技大学出版社

内 容 简 介

本书涵盖计算机图形学基础所对应的主要实验,以期让读者系统掌握计算机图形学基础知识,提高读者的实践能力和创新能力。为改善学习效果,本书基于主流图形渲染管线的流程,设计了整体式实验框架,在其中添加当前课程内容即可得到一个可运行的图形程序,实现理论与实践的密切关联,以有效激发读者的实验兴趣并提升其实践能力,有利于随堂实践教学的开展。

本书主要内容包括实验环境介绍、实验程序框架介绍、图元扫描转换、二维几何变换、三维几何变换、直线段和多边形裁剪、图元填充、几何图元消隐、三维图形的表示和加载、光照计算、纹理映射等。本书还为每个实验提供了线上教学资源(包括详细的文本阐述及示例程序演示等内容),方便读者更加自由地开展学习。书中同时部署了课外拓展性实验,可帮助读者自我评估实验相关图形技术的掌握情况,激发其探索能力和创新能力。

本书面向本科阶段的计算机图形学课程教学,可作为计算机图形学的入门教材,也可配合市面上大部分计算机图形学教材使用。

图书在版编目(CIP)数据

线上线下混合式计算机图形学基础实验教程/潘万彬等编著. —西安:西安电子科技大学出版社,2021.8 (2022.7 重印)
ISBN 978−7−5606−6147−6

Ⅰ. ① 线… Ⅱ. ① 潘… Ⅲ. ① 计算机图形学—高等学校—教材 Ⅳ. ① TP391.411

中国版本图书馆 CIP 数据核字(2021)第 149465 号

策　　划　陈　婷
责任编辑　祝婷婷　陈　婷
出版发行　西安电子科技大学出版社(西安市太白南路 2 号)
电　　话　(029)88202421　88201467　　邮　编　710071
网　　址　www.xduph.com　　电子邮箱　xdupfxb001@163.com
经　　销　新华书店
印刷单位　陕西天意印务有限责任公司
版　　次　2021 年 8 月第 1 版　　2022 年 7 月第 2 次印刷
开　　本　787 毫米×960 毫米　1/16　印　张　9.25
字　　数　180 千字
印　　数　501～1500 册
定　　价　30.00 元

ISBN 978−7−5606−6147−6/TP

XDUP 6449001−2

如有印装问题可调换

前　言

👉 编写目的

随着我国制造 2025 战略的稳步推进，计算机图形学在新兴产业(包括与大数据可视化、数字孪生、影视动漫、虚拟/增强现实等技术关联的产业)中扮演着越来越重要的角色，并逐步成为推动我国数字经济、智能制造等持续发展不可或缺的一项技术。为了使读者能够适应上述发展需求，帮助读者系统掌握计算机图形学基础的核心知识，我们面向计算机图形学基础实验编写了本书。

期望本书的编写和出版能够提高更多读者对计算机图形学课程的学习效果，培养和增强他们在计算机图形学上的实践能力、探索能力和创新能力；同时，也希望为高校新工科专业(如数字媒体技术、数据科学与大数据技术等)课程体系的建设提供有益的借鉴。

👉 核心内容、特点

本书侧重计算机图形学基础的主要实验，包括图元扫描转换、二维几何变换、三维几何变换、直线段和多边形裁剪、图元填充、几何图元消隐、三维图形的表示和加载、光照计算、纹理映射等。同时，本书针对每个实验提供了下列内容：

(1) 线上教学，以视频、电子版学习资源等形式，提供实验必要的基础知识、实践操作指导等，以便读者顺利开展实验。同时，在纸质实验教材中嵌入线上教学链接二维码，以便读者通过扫描二维码开展在线学习。

(2) 示例程序(包括关键代码)，以便读者开展实践教学或自学。

(3) 课外拓展性实验，以便读者自我评估实验相关图形技术的掌握情况，激发其探索能力和创新能力。

本书借助"互联网+"科技手段，对计算机图形学基础实验开展线上线下混合式教学，以期让读者深入理解和掌握计算机图形学的基础知识及图形编程的关键问题，为深入学习高阶计算机图形学内容打下扎实的基础。

👉 适用范围

本书可配合线上教学内容作为计算机图形学的入门教材，同时也可配合市面上大部分计算机图形学教材使用。

✍ 作者分工

潘万彬为本书的总负责人，王毅刚承担实验内容的总体设计，曹伟娟参与本书涉及的视频资源建设规划，王若兰、洪雨洁、黄婷婷、徐纯芝和董琛琛主要负责代码和文档的整理以及视频资源和插图的制作。

✍ 阅读本书所需要的预备知识

计算机图形学涉及大量的图形算法、数据结构等计算机知识，且与数学、物理紧密关联。因此，在阅读本书之前，读者需要具备一定的计算机编程知识和基本的数据结构知识，如指针、数组、链表等。同时，在计算机图形学算法中还涉及一些数学方法，读者需要具备相关的数学知识，如几何、三角、线性代数等。在使用过程中，读者可根据本书提供的参考文献，熟悉相关的 OpenGL 基础知识。有了上述的预备知识，即使读者在阅读本书之前对计算机图形学的理论知识了解较少，也仍然可以理解和掌握本书中所讨论的绝大多数内容。

✍ 如何获取本书示例程序的源代码和相关视频

读者可以通过扫描书中二维码或从西安电子科技大学出版社网站(www.xduph.com)下载，获取本书示例程序的源代码和实践指导教学视频。

编著者

2021 年 5 月

目　　录

第1章　实验环境介绍 1
1.1　实验平台 1
1.2　实验项目创建 1
第2章　实验程序框架介绍 6
第3章　图元扫描转换 10
3.1　实验内容简述和实验目标 10
3.2　扫描转换直线段(DDA) 10
3.3　基于 Bresenham 算法绘制直线段 12
3.4　基于 Bresenham 算法绘制圆 14
3.5　课外拓展性实验 16
第4章　二维几何变换 17
4.1　实验内容简述和实验目标 17
4.2　二维基本几何变换 17
4.2.1　非齐次平移变换 17
4.2.2　非齐次缩放变换 19
4.2.3　非齐次旋转变换 20
4.2.4　非齐次对称变换 21
4.2.5　非齐次错切变换 23
4.2.6　齐次平移变换 25
4.2.7　齐次缩放变换 28
4.2.8　齐次旋转变换 29
4.2.9　齐次对称变换 31
4.2.10　齐次错切变换 34
4.3　二维复合几何变换 37
4.3.1　绕任意点旋转变换 37
4.3.2　相对任意点缩放变换 38
4.4　课外拓展性实验 40
第5章　三维几何变换 41
5.1　实验内容简述和实验目标 41

5.2 三维基本几何变换 41
5.2.1 三维平移变换 41
5.2.2 三维旋转变换 44
5.2.3 三维缩放变换 49
5.2.4 三维错切变换 51
5.2.5 三维对称变换 56
5.3 三维复合几何变换 67
5.3.1 三维图形绕空间任意轴旋转 67
5.3.2 三维图形相对任意点缩放 69
5.4 三维几何变换综合示例 70
5.5 课外拓展性实验 71

第6章 直线段和多边形裁剪 72
6.1 实验内容简述和实验目标 72
6.2 Cohen-Sutherland 直线段裁剪 72
6.3 Sutherland-Hodgman 多边形裁剪 77
6.4 Weiler-Atherton 多边形裁剪 82
6.5 课外拓展性实验 87

第7章 图元填充 88
7.1 实验内容简述和实验目标 88
7.2 扫描线填充任意多边形 88
7.3 课外拓展性实验 94

第8章 几何图元消隐 95
8.1 实验内容简述和实验目标 95
8.2 z 缓冲消隐 95
8.3 背面剔除消隐 98
8.4 课外拓展性实验 99

第9章 三维图形的表示和加载 100
9.1 实验内容简述和实验目标 100
9.2 OBJ 格式的多边形表示模型 100
9.2.1 OBJ 格式文件含义 100
9.2.2 OBJ 模型内容的读入与相关数据结构的装配 102
9.2.3 OBJ 模型的绘制 105
9.3 课外拓展性实验 107

第 10 章 光照计算 108
10.1 实验内容简述和实验目标 108
10.2 直接光照计算 108
10.2.1 环境光 108
10.2.2 漫反射光 109
10.2.3 镜面反射光 110
10.2.4 Phong 光照模型 111
10.2.5 Blinn-Phong 光照模型 113
10.3 明暗过渡计算 114
10.3.1 Flat 明暗过渡 114
10.3.2 Gouraud 明暗过渡 116
10.3.3 Phong 明暗过渡 118
10.4 全局光照之光线跟踪算法 120
10.4.1 光线跟踪中的求交计算 121
10.4.2 反射光线方向的计算 121
10.4.3 折射光线方向的计算 122
10.5 课外拓展性实验 123

第 11 章 纹理映射 124
11.1 实验内容简述和实验目标 124
11.2 纹理映射 124
11.2.1 创建纹理 124
11.2.2 设置纹理参数 125
11.2.3 设定映射方式 125
11.2.4 映射纹理坐标与几何坐标 126
11.2.5 设置纹理的重复度 128
11.3 Shadow Mapping 阴影生成 129
11.4 课外拓展性实验 138

参考文献 139

第1章 实验环境介绍

本书结合 OpenGL 渲染管线开展实验。为了让学生深入透彻理解图形学原理,本书绝大部分实验采用 C/C++实现,仅需用到 OpenGL 的绘制顶点、顶点着色及绑定纹理等基本功能。

1.1 实验平台

编程综合平台:Windows 系统,Microsoft Visual C++ 2003 及以上。
图形库:OpenGL 1.0 版本及以上。

1.2 实验项目创建

为了顺利开展实验项目,可以按照以下步骤来完成图形程序环境的创建。此处结合 Visual Studio 2019 进行具体介绍。

第一步:打开 Visual Studio 2019,选择创建新项目,如图 1-1 所示。

环境配置文件包

图 1-1　创建新项目操作界面

第二步：选择编程语言 C++，并选择"空项目"，点击"下一步"按钮，如图 1-2 所示。

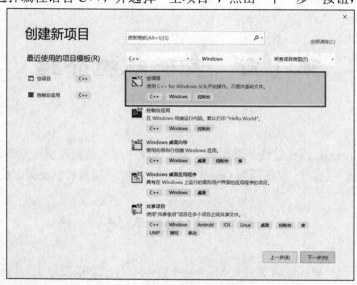

图 1-2　应用类型和语言选择

第三步：填写项目名称，如 CG，并确定路径，如 C:\Users\小洪同学\Desktop\，之后点击"创建"按钮，如图 1-3 所示。

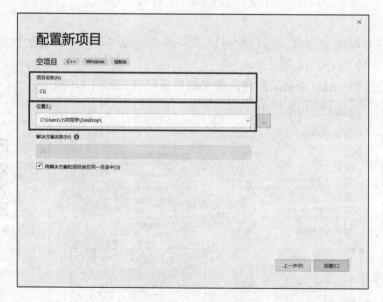

图 1-3　项目创建示意图

第四步：右键点击"源文件"，在弹出的菜单中选择"添加"→"新建项"，之后选择

第 1 章　实验环境介绍

"C++文件",并命名为 main.cpp,如图 1-4 和图 1-5 所示。

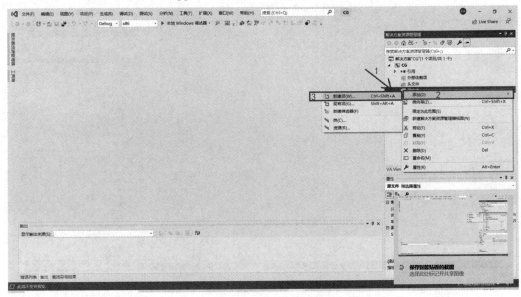

图 1-4　源文件添加(1)

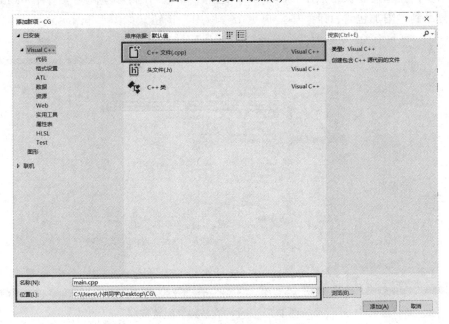

图 1-5　源文件添加(2)

第五步:为了在编译和链接时能够使用 OpenGL 的库文件,需要对刚创建的项目进行环境配置，具体操作为右键点击项目"CG"，然后选择"属性"，如图1-6所示。

图1-6 开始配置环境

第六步：在属性面板中点击"C/C++"，选择"常规"，将环境配置文件中include文件的地址复制到"附加包含目录"中，再点击"应用"按钮，如图1-7所示。

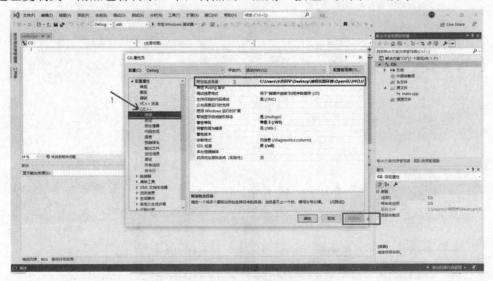

图1-7 头文件存放路径配置

第七步：在属性面板中点击"链接器"，选择"常规"，将环境配置文件中lib文件的地址复制到"附加库目录"中，再点击"确定"或"应用"按钮，如图1-8所示。

第 1 章　实验环境介绍

图 1-8　链接库存放路径配置

第八步：在属性面板中点击"链接器"，选择"输入"，将"opengl32.lib;glu32.lib;glut32.lib"添加到"附加依赖项"中，再点击"确定"按钮，如图 1-9 所示。

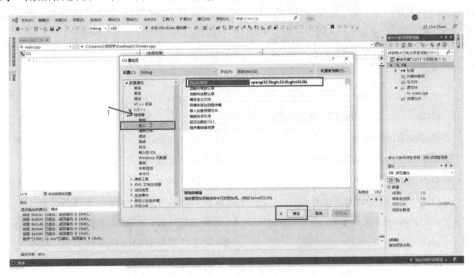

图 1-9　链接库附加依赖项添加

第九步：环境配置结束，在源文件中添加框架代码，即可进行后续操作。

第 2 章 实验程序框架介绍

为了便于后期整体式嵌入编程，使读者将注意力重点放在每一个实验内容的实现上，本书对所有实验使用了统一的程序框架。此处结合通用的 OpenGL 环境设置进行介绍。框架共涉及 1 个头文件 GLUT.H 和 5 个主要函数(main、reshape、display、keyboard 和 mouse)。下面对 5 个主要函数进行解释说明。

实验程序框架介绍

1. main 函数

本书以 main 函数作为所有图形程序实验的主调函数和入口。该函数主要用于初始化图形程序窗口的位置、大小，图形程序的显示方式，以及设置各类事件对应的回调函数，如窗口尺寸变换、鼠标键盘交互等。具体代码如下：

```
int main(int argc, char** argv)
{
    glutInit(&argc, argv);                              //初始化 GLUT 库
    glutInitDisplayMode(GLUT_SINGLE | GLUT_RGB);        //初始化显示模式
    glutInitWindowSize(500, 500);                       //设置窗口的大小
    glutInitWindowPosition(50, 50);                     //设置窗口的位置
    glutCreateWindow("HelloWorld");                     //创建窗口
    glutReshapeFunc(reshape);                           //设置窗口尺寸变换时对应的回调函数
    glutDisplayFunc(display);                           //设置绘制时对应的回调函数
    glutKeyboardFunc(keyboard);                         //设置键盘交互时对应的回调函数
    glutMouseFunc(mouse);                               //设置鼠标交互时对应的回调函数
    glutMainLoop();                                     //启动图形程序
    return 0;
}
```

2. reshape 函数

为使图形程序能响应窗口尺寸的改变，及时调整图形程序的投影方式或相应的投影参

数、modelview 变换方式(即三维空间中的模型或视点变换)，以及视口变换方式(即关联投影面到窗口的变换)，本书定义了 reshape 函数。该函数内部涉及的 OpenGL 函数可参考 OpenGL 相关资料。具体代码如下：

```
void reshape(int w, int h)
{
    glViewport(0, 0, (GLsizei)w, (GLsizei)h);        //视口变换
    glMatrixMode(GL_PROJECTION);                      //设置投影模式以及视景体大小
    glLoadIdentity();                                 //重置投影变换矩阵
    if (w <= h)                                       //平行投影
        glOrtho(-16.0, 16.0, -16.0 * (GLfloat)h / (GLfloat)w,
                16.0 * (GLfloat)h / (GLfloat)w, -ZVALUE, ZVALUE);
    else
        glOrtho(-16.0 * (GLfloat)h / (GLfloat)w, 16.0 * (GLfloat)h / (GLfloat)w, -16.0, 16.0,
                -ZVALUE, ZVALUE);
    glMatrixMode(GL_MODELVIEW);                       //开始设置模型视点变换矩阵
    glLoadIdentity();                                 //重置模型视点变换矩阵
}
```

3. display 函数

display 函数主要用于对图形场景内容、坐标系进行绘制，以及对窗口的背景颜色进行设置等。具体代码如下：

```
void display(void)
{
    glClearColor(1.f, 1.f, 1.f, 0.f);                //设置窗口背景色
    glClear(GL_COLOR_BUFFER_BIT);                    //用当前背景色填充窗口
    draw_coordinate();                               //绘制坐标系，自定义函数
    glColor3f(0, 0, 1);                              //设置当前的绘图颜色
    glBegin(GL_QUADS);                               //绘制四边形
        glVertex2f(0, 0);
        glVertex2f(0, -3);
        glVertex2f(10, -3);
        glVertex2f(10, 0);
    glEnd();
    glFlush();                                       //清空 OpenGL 命令缓冲区，执行 OpenGL 命令
```

}

4. keyboard 函数

keyboard 函数用于响应键盘交互事件，通过键盘与图形程序或其中的图形元素进行交互。具体代码如下：

```
void keyboard(unsigned char key, int x, int y)
{
    switch (key)
    {
     case 'x':
     case 'X':
     {
        //此处可以调用函数改变图形空间中元素的位置或姿态，实现键盘交互
        glutPostRedisplay();    //重新绘制，会自动调用 display 函数
        break;
     }
     case 27:
        exit(0);
        break;
    }
}
```

5. mouse 函数

mouse 函数用于响应鼠标交互事件，通过鼠标左、右键或中键与图形程序或其中的图形元素进行交互。具体代码如下：

```
void mouse(int button, int state, int x, int y)
{
    switch (button)
    {
     case GLUT_LEFT_BUTTON:
        if (state == GLUT_DOWN)     //左键点击
        {
            //此处可以调用函数改变图形空间中元素的位置或姿态，实现鼠标交互
            glutPostRedisplay();
        }
```

            break;
        ...
    case GLUT_RIGHT_BUTTON:
        ...
    default:
        break;
    }
}

可以用本框架代码验证第 1 章的环境配置是否正确，正确运行的效果如图 2-1 所示。

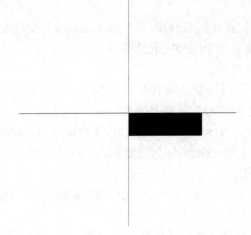

图 2-1 效果图

第 3 章 图元扫描转换

3.1 实验内容简述和实验目标

基本实验内容包括:扫描转换直线段、基于 Bresenham 算法绘制直线段、基于 Bresenham 算法绘制圆。同时,配备了一个课外拓展性实验。

完成本实验后,读者能够:

(1) 描述光栅化的原因和原理(布鲁姆知识模型:记忆和理解);

(2) 列举常见的光栅化绘制现象(布鲁姆知识模型:理解);

(3) 复述 DDA、中点、Bresenham 算法流程(布鲁姆知识模型:记忆);

(4) 推导 DDA、中点、Bresenham 算法绘制二次曲线的递推公式,并能够算法化(布鲁姆知识模型:理解和应用);

(5) 基于 OpenGL,采用中点、Bresenham 算法编程绘制二次曲线(布鲁姆知识模型:应用);

(6) 检查并发现图元光栅化绘制效果存在的问题,判断原因,找出对策并改善(布鲁姆知识模型:应用、分析、评价)。

3.2 扫描转换直线段(DDA)

给定直线方程 $y = k \times x + b$ 上两点 (x_0, y_0) 和 (x_1, y_1),绘制上述两点之间的直线段。

$$k = \frac{dy}{dx} = \frac{y_1 - y_0}{x_1 - x_0} \quad (x_1 \neq x_0)$$
$$y_{i+1} = y_i + k$$
(3.1)

不失一般性地,此处假定斜率 k 的绝对值小于等于 1,且 $x_1 > x_0$。基于 DDA 递推公式 (3.1),从点 (x_0, y_0) 开始,朝着 (x_1, y_1) 所在位置前进一个像素。期间,直线段上下一点坐标

的 x 值等于当前点坐标的 x 值加 1，并根据公式(3.1)计算出直线段上下一点坐标的 y 值。因为像素的坐标是整数，所以还需要对上述 x 和 y 值进行取整处理，以得到逼近所给直线段的下一像素坐标(x_i, y_i)。对应算法步骤为：

(1) 输入直线段的两端点 $p_0(x_0, y_0)$ 和 $p_1(x_1, y_1)$。

(2) 计算公式(3.1)中的初始值 k。

(3) 令直线段上的点为(x_f, y_f)，初始为(x_0, y_0)。

(4) 当 $x_f <= x_1$ 时，迭代以下两步骤，否则结束

① 将(x_f, y_f)取整为(x_i, y_i)，并进行绘制。

② 计算直线段上的下一点，即 $x_f = x_f + 1$，并根据公式(3.1)计算下一点的 y_f。

1. 关键函数代码实现

代码如下：

```
void DDA(int x0, int y0, int x1, int y1)
{
    float dy = y1 - y0;
    float dx = x1 - x0;
    float k = dy / dx;
    float xf = x0;
    float yf = y0;
    while (xf <= x1)
    {
        //对当前点进行舍入，获取像素点坐标
        int xi = (floor)(xf + 0.5);
        int yi = (floor)(yf + 0.5);
        glVertex2i(xr,yr);
        //计算直线段上下一个点
        xf = xf + 1;
        yf = yf + k;
    }
}
```

2. 案例效果

用 DDA 方法绘制一条直线段，直线段的两端点如图 3-1(a)所示，坐标分别为(-200，-200)和(200，200)，最终效果如图 3-1(b)所示。

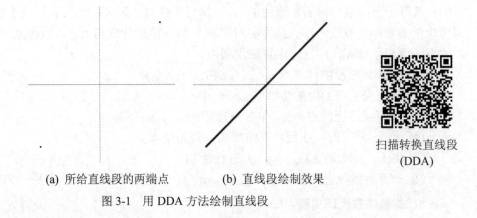

(a) 所给直线段的两端点 　　　　(b) 直线段绘制效果

图 3-1　用 DDA 方法绘制直线段

3.3　基于 Bresenham 算法绘制直线段

假设当前像素点坐标为$(x_i, y_{i,r})$，且当前直线段的斜率小于等于 1，则逼近直线段的下一个像素点坐标$(x_{i+1}, y_{i+1,r})$的具体取值规则如下：$x_{i+1} = x_i + 1$；$y_{i+1,r}$的值由公式(3.2)给出。这里，d_i为$y_{i+1,r}$取值的判别公式，由其决定下一像素点为$(x_{i+1}, y_{i,r})$还是$(x_{i+1}, y_{i,r} + 1)$。此处，令$d_i = y_{i+1} - y_{i,r} - 0.5$，其中$y_{i+1}$为当前直线段与$x = x_{i+1}$相交时的真实$y$坐标。为此，当$d_i \geqslant 0$时，说明当前直线段的下一实际交点$(x_{i+1}, y_{i+1})$离$(x_{i+1}, y_{i,r} + 1)$更近，即$y_{i+1,r}$更适合为$y_{i,r} + 1$；否则，$y_{i+1,r}$更适合取值$y_{i,r}$。同时，判别式的递推公式为$d_{i+1} = y_{i+2} - y_{i+1,r} - 0.5 = y_{i+1} + k - y_{i+1,r} - 0.5$，也取决于$y_{i+1,r}$的取值，如公式(3.3)所示。此处，$k$为当前直线段的斜率$(d_y \div d_x)$，为此，$d_i$的初始值为$k - 0.5$。

$$y_{i+1,r} = \begin{cases} y_{i,r} + 1 & \text{当} d_i \geqslant 0 \text{时} \\ y_{i,r} & \text{其他情况} \end{cases} \quad (3.2)$$

$$d_{i+1} = \begin{cases} d_i + k - 1 & \text{当} d_i >= 0 \text{时} \\ d_i + k & \text{其他情况} \end{cases} \quad (3.3)$$

为了提高算法的效率，对上述公式实施整数计算。为此，对判别式乘以$2 \times dx$，此时公式(3.3)更新为公式(3.4)，且初始d_i为$2 \times dy - dx$。

$$d_{i+1} = \begin{cases} d_i + 2 \times (d_y - d_x) & \text{当} d_i >= 0 \text{时} \\ d_i + 2 \times d_y & \text{其他情况} \end{cases} \quad (3.4)$$

对应算法步骤如下：

(1) 输入直线的两端点 $p_0(x_0, y_0)$ 和 $p_1(x_1, y_1)$。

(2) 计算初始值 d_y，d_x，d。

(3) 绘制点 (x_f, y_f)，初始值为 (x_0, y_0)。

(4) 判断 d 的符号。若 $d >= 0$，则下一个 (x, y) 更新为 $(x+1, y+1)$，同时将 d 更新为 $d + 2 \times (d_y - d_x)$；否则，将下一个 (x, y) 更新为 $(x+1, y)$，d 更新为 $d + 2 \times d_y$。

(5) 当 $x_f < x_1$ 时，迭代步骤(3)和(4)。否则结束。

1. 关键函数代码实现

代码如下：

```
void Bresenham(int x0, int y0, int x1, int y1)
{
    int dy = y1 - y0;
    int dx = x1 - x0;
    int xf = x0;
    int yf = y0;
    int d = 2 * dy - dx;
    while (xf < x1)
    {
        glVertex2d(xf, yf);
        if (d >= 0)
        {
            yf++;
            xf++;
            d = d + 2 * (dy - dx);
        }
        else
        {
            xf++;
            d = d + 2 * dy;
        }
    }
}
```

2. 案例效果

用 Bresenham 算法绘制一条直线段，直线段的两端点如图 3-2(a)所示，坐标分别为(0, 0)和(100, 100)，最终效果如图 3-2(b)所示。

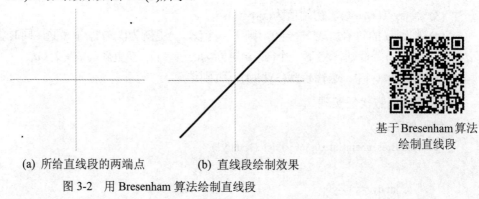

基于Bresenham算法绘制直线段

(a) 所给直线段的两端点　　　(b) 直线段绘制效果

图 3-2　用 Bresenham 算法绘制直线段

3.4　基于 Bresenham 算法绘制圆

基于 Bresenham 算法绘制圆的算法的主要步骤如下：

(1) 输入圆的半径 r，默认圆心为(0, 0)。
(2) 计算初始值 x、y，判别式 $d=3-2r$。
(3) 根据圆的对称性绘制点(x, y)，$(-x, y)$，$(x, -y)$，$(-x, -y)$，(y, x)，$(-y, x)$，$(y, -x)$，$(-y, -x)$。
(4) 根据 d 的符号确定下一点 y 的取值。如果 $d>0$，则下一个(x, y)更新为$(x+1, y-1)$，同时将判别式 d 更新为 $d+4(x-y)+10$；否则下一个(x, y)更新为$(x+1, y)$，将判别式 d 更新为 $d+4x+6$。
(5) 迭代步骤(3)和(4)，直至圆绘制结束。

1. 关键函数代码实现

代码如下：

```
//默认绘制圆的圆心在原点，r 为半径
void Circle(int r)
{
    int x, y;
    x = 0, y = r;
    int d = 3 - 2 * r;
```

```
while (x <= y)
{
    glVertex2d(x, y);
    glVertex2d(-x, y);
    glVertex2d(x, -y);
    glVertex2d(y, x);
    glVertex2d(-x, -y);
    glVertex2d(-y, -x);
    glVertex2d(y, -x);
    glVertex2d(-y, x);
    if (d > 0)
    {
        d += 4 * (x - y) + 10;
        y--;
    }
    else
    {
        d += 4 * x + 6;
    }
    x++;
}
```

2. 案例效果

用 Bresenham 算法绘制一个圆心在(0,0)、半径为 200 的圆,最终效果如图 3-3 所示。

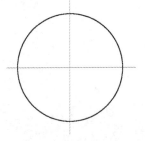

基于 Bresenham 算法绘制圆

图 3-3 用 Bresenham 算法绘制圆的最终效果图

3.5 课外拓展性实验

用 Bresenham 算法绘制一个椭圆(见图 3-4),椭圆的长、短半轴分别为 200、100。

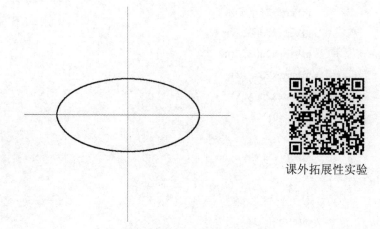

课外拓展性实验

图 3-4 用 Bresenham 算法绘制椭圆的最终效果图

第4章 二维几何变换

4.1 实验内容简述和实验目标

基本实验内容包括：二维基本几何变换(非齐次平移变换、非齐次缩放变换、非齐次旋转变换、非齐次对称变换、非齐次错切变换、齐次平移变换、齐次缩放变换、齐次旋转变换、齐次对称变换、齐次错切变换)和二维复合几何变换(绕任意点旋转变换、相对任意点缩放变换)。同时，配备了一个课外拓展性实验。

完成本实验后，读者能够：

(1) 列举常见二维图形的基本几何变换，并能够描述它们的概念(布鲁姆知识模型：记忆)；

(2) 解释二维几何图形产生几何变换的原理(布鲁姆知识模型：理解)；

(3) 描述出二维图形基本几何变换代数表达式在齐次和非齐次坐标下的推导过程，并牢记相应的推导结果(布鲁姆知识模型：记忆和理解)；

(4) 归纳二维图形基本几何变换采用齐次坐标系下矩阵表示的原因(布鲁姆知识模型：分析)；

(5) 分解任意给定的二维图形复合几何变换为二维图形基本几何变换序列，并给出相应的矩阵复合表示(布鲁姆知识模型：应用和分析)；

(6) 运用矩阵与二维图形基本几何变换之间的对应关系，预测任意给定的矩阵复合所对应的几何图形发生的复合几何变换(布鲁姆知识模型：应用)；

(7) 编程实现——通过鼠标、键盘交互，展现给定二维图形的基本几何变换和任意复合几何变换(布鲁姆知识模型：应用)。

4.2 二维基本几何变换

4.2.1 非齐次平移变换

二维图形非齐次平移的参数化表达式如下：

$$\begin{cases} x' = x + t_x \\ y' = y + t_y \end{cases} \tag{4.1}$$

其中，t_x、t_y 分别表示物体在 x、y 轴方向上的平移量。

1. 关键数据结构

自定义如下数据结构用以表示二维非齐次坐标下的顶点和向量。该数据结构应用于本章所有的非齐次二维几何变换实验。

```
struct my_v_inhomogeneous
{
    float x;
    float y;
};
```

2. 变换函数代码实现

代码如下：

```
/*********************************
 * 非齐次平移变换
 * polygon：几何图形；polygon_vertex_count：几何图形的顶点数量；
 * tx、ty：x 轴、y 轴方向上的平移量
 * 将原坐标的 x、y 值分别加上平移量 tx、ty 来实现平移
 **********************************/
void my_traslate_inhomogeneous(struct my_v_inhomogeneous* polygon,
                    int polygon_vertex_count, int tx, int ty)
{
    for (int vIndex = 0; vIndex < polygon_vertex_count; vIndex++)
    {
        polygon[vIndex].x += tx;
        polygon[vIndex].y += ty;
    }
}
```

3. 案例效果

采用非齐次平移变换方式，对如图 4-1(a)所示的中心在原点的三角形，向 x 轴正方向平移 40 个单位，向 y 轴正方向平移 50 个单位，最终效果如图 4-1(b)所示。

第 4 章 二维几何变换

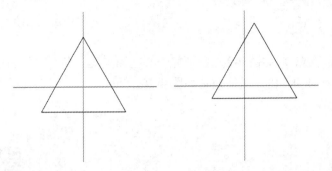

非齐次平移变换

(a) 初始图形　　　　(b) 平移后的图形

图 4-1　采用非齐次平移变换方式示意图

4.2.2　非齐次缩放变换

二维图形非齐次缩放的参数化表达式如下：

$$\begin{cases} x' = s_x \times x \\ y' = s_y \times y \end{cases} \tag{4.2}$$

其中，s_x、s_y 分别表示物体在 x、y 轴方向上的缩放量。

1. 变换函数代码实现

代码如下：

```
/*******************************
 * 非齐次缩放变换
 * polygon：几何图形；polygon_vertex_count：几何图形的顶点数量；
 * sx、sy：x 轴、y 轴方向上的缩放系数
 * 将原本坐标的 x、y 值分别乘上 x 轴、y 轴方向上的缩放系数来实现缩放
 *******************************/
void my_scale_inhomogeneous(struct my_v_inhomogeneous* polygon,
                            int polygon_vertex_count, float sx, float sy)
{
    for (int vIndex = 0; vIndex < polygon_vertex_count; vIndex++)
    {
        polygon[vIndex].x *= sx;
        polygon[vIndex].y *= sy;
    }
```

}

2. 案例效果

采用非齐次缩放变换方式,对如图 4-2(a)所示的中心在原点的三角形,在 x 轴方向上缩放 1.5 倍,在 y 轴方向上缩放 0.5 倍,最终效果如图 4-2(b)所示。

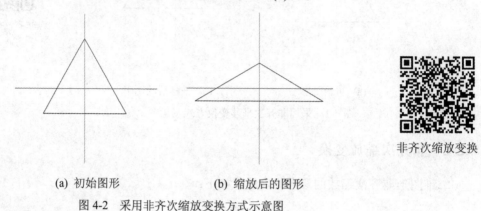

非齐次缩放变换

(a) 初始图形　　　　　(b) 缩放后的图形

图 4-2　采用非齐次缩放变换方式示意图

4.2.3　非齐次旋转变换

二维图形非齐次旋转变换的参数化表达式如下:

$$\begin{cases} x' = \cos\theta \times x - \sin\theta \times y \\ y' = \sin\theta \times x + \cos\theta \times y \end{cases} \quad (4.3)$$

其中,θ 表示物体逆时针绕点旋转的角度。

1. 变换函数代码实现

代码如下:

```
/*******************************
* 非齐次旋转变换
* polygon：几何图形；polygon_vertex_count：几何图形的顶点数量；
* angle：几何图形的旋转角度,正值逆时针旋转,负值顺时针旋转
* 根据式(4.3)得到的变换后坐标与变换前坐标的关系编写代码
*******************************/
void my_rotate_inhomogeneous(struct my_v_inhomogeneous* polygon,
                             int polygon_vertex_count, float angle)
{
    angle = angle * 3.14 / 180;
```

```
float x, y;
for (int vIndex = 0; vIndex < polygon_vertex_count; vIndex++)
{
    x = polygon[vIndex].x;
    y = polygon[vIndex].y;
    polygon[vIndex].x = x * cos(angle) - y * sin(angle);
    polygon[vIndex].y = x * sin(angle) + y * cos(angle);
}
```

2. 案例效果

采用非齐次旋转变换方式，对如图 4-3(a)所示的中心在原点的三角形逆时针旋转 50°，最终效果如图 4-3(b)所示。

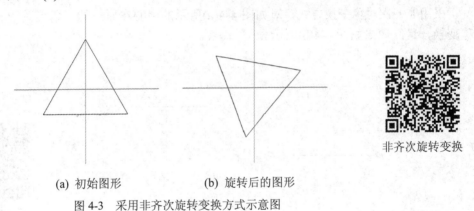

非齐次旋转变换

(a) 初始图形　　　　(b) 旋转后的图形

图 4-3　采用非齐次旋转变换方式示意图

4.2.4　非齐次对称变换

1. 关于 x 轴的非齐次对称

二维图形关于 x 轴的非齐次对称变换的参数化表达式如下：

$$\begin{cases} x' = x \\ y' = -y \end{cases} \tag{4.4}$$

1) 变换函数代码实现

代码如下：

```
/*******************************
* 关于 x 轴的非齐次对称变换
```

```
* polygon：几何图形；polygon_vertex_count：几何图形的顶点数量
* 根据式(4.4)得到的变换后坐标与变换前坐标的关系编写代码
**********************************/
void my_reflection_x_inhomogeneous(struct my_v_inhomogeneous* polygon,
                                    int polygon_vertex_count)
{
    for (int vIndex = 0; vIndex < polygon_vertex_count; vIndex++)
    {
        polygon[vIndex].y = -polygon[vIndex].y;
    }
}
```

2) 案例效果

采用非齐次对称变换方式，对如图 4-4(a)所示的中心在原点的三角形关于 x 轴进行对称变换，最终效果如图 4-4(b)所示。

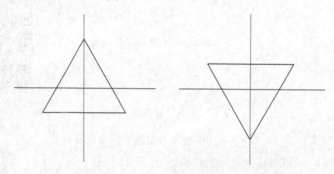

(a) 初始图形　　(b) 对称后的图形

关于 x 轴的非齐次对称变换

图 4-4　关于 x 轴的非齐次对称变换方式示意图

2. 关于 y 轴的非齐次对称

二维图形关于 y 轴的非齐次对称变换的参数化表达式如下：

$$\begin{cases} x' = -x \\ y' = y \end{cases} \quad (4.5)$$

1) 变换函数代码实现

代码如下：

```
/**********************************
* 关于 y 轴的非齐次对称变换
```

```
* polygon：几何图形；polygon_vertex_count：几何图形的顶点数量
* 根据式(4.5)得到的变换后坐标与变换前坐标的关系编写代码
******************************************/
void my_reflection_y_inhomogeneous(struct my_v_inhomogeneous* polygon,
                                   int polygon_vertex_count)
{
    for (int vIndex = 0; vIndex < polygon_vertex_count; vIndex++)
    {
        polygon[vIndex].x = -polygon[vIndex].x;
    }
}
```

2) 案例效果

采用非齐次对称变换方式，对如图 4-5(a)所示的中心在原点的三角形关于 y 轴进行对称变换，最终效果如图 4-5(b)所示。

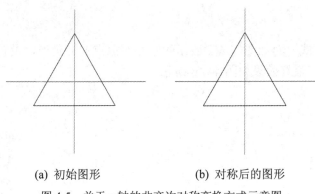

(a) 初始图形　　　　(b) 对称后的图形

图 4-5　关于 y 轴的非齐次对称变换方式示意图

关于 y 轴的非齐次对称变换

4.2.5　非齐次错切变换

1. 在 x 轴方向上的非齐次错切

二维图形在 x 轴方向上的非齐次错切变换的参数化表达式如下：

$$\begin{cases} x' = x + \mathrm{sh}_x \times y \\ y' = y \end{cases} \quad (4.6)$$

错切变换过程中，y 坐标值保持不变，x 坐标值根据 y 坐标以 sh_x 为切变程度发生线性变化。

1) 变换函数代码实现

代码如下:

```
/*************************************
* 在 x 轴方向上的非齐次错切
* polygon：几何图形；polygon_vertex_count：几何图形的顶点数量；shx：切变程度
* 根据式(4.6)得到的变换后坐标与变换前坐标的关系编写代码
*************************************/
void my_shear_x_inhomogeneous(struct my_v_inhomogeneous* polygon,
                              int polygon_vertex_count, float shx)
{
    for (int vIndex = 0; vIndex < polygon_vertex_count; vIndex++)
    {
        polygon[vIndex].x = polygon[vIndex].x + shx * polygon[vIndex].y;
    }
}
```

2) 案例效果

采用非齐次错切变换方式，对如图 4-6(a)所示的中心在原点的三角形，设置相对于 x 轴方向上的错切参数为 1.2，最终效果如图 4-6(b)所示。

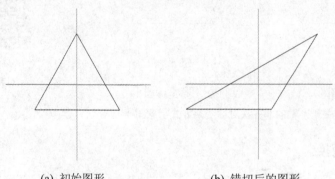

(a) 初始图形　　　　　(b) 错切后的图形

图 4-6　在 x 轴方向上的非齐次错切变换方式示意图

在 x 轴方向上的非齐次错切

2. 在 y 轴方向上的非齐次错切

二维图形在 y 轴方向上的非齐次错切变换的参数化表达式如下：

$$\begin{cases} x' = x \\ y' = \mathrm{sh}_y \times x + y \end{cases} \tag{4.7}$$

错切变换过程中，x 坐标值保持不变，y 坐标值根据 x 坐标以 sh_y 为切变程度发生线性变化。

1) 变换函数代码实现

代码如下：

```
/*****************************************
* 相对于 y 轴方向的非齐次错切
* polygon：几何图形；polygon_vertex_count：几何图形的顶点数量；shy：切变程度
* 根据式(4.7)得到的变换后坐标与变换前坐标的关系编写代码
*****************************************/
void my_shear_y_inhomogeneous(struct my_v_inhomogeneous* polygon,
                              int polygon_vertex_count, float shy)
{
    for (int vIndex = 0; vIndex < polygon_vertex_count; vIndex++)
    {
        polygon[vIndex].y = shy * polygon[vIndex].x + polygon[vIndex].y;
    }
}
```

2) 案例效果

采用非齐次错切变换方式，对如图 4-7(a)所示的中心在原点的三角形，设置相对于 y 轴方向上的错切参数为 0.8，最终效果如图 4-7(b)所示。

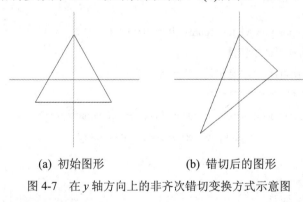

(a) 初始图形　　　(b) 错切后的图形

在 y 轴方向上的非齐次错切

图 4-7　在 y 轴方向上的非齐次错切变换方式示意图

4.2.6　齐次平移变换

二维图形齐次平移的参数化表达式如下：

$$\begin{cases} x' = x + t_x \\ y' = y + t_y \end{cases} \qquad (4.8)$$

其中，t_x、t_y 分别表示物体在 x、y 轴方向上的平移量。

对应的齐次平移变换的矩阵乘积形式如下：

$$\begin{pmatrix} x' \\ y' \\ 1 \end{pmatrix} = \begin{pmatrix} 1 & 0 & t_x \\ 0 & 1 & t_y \\ 0 & 0 & 1 \end{pmatrix} \begin{pmatrix} x \\ y \\ 1 \end{pmatrix} \qquad (4.9)$$

1. 关键数据结构

自定义如下数据结构用以表示二维齐次坐标下的顶点和向量。该数据结构应用于本章所有齐次二维几何变换实验。

```
struct my_v_homogeneous
{
    float x;
    float y;
    int ratio;
};
```

2. 变换函数及关键相关代码实现

代码如下：

```
/*******************************
* 齐次平移变换
* polygon：几何图形；polygon_vertex_count：几何图形的顶点数量；
* tx、ty：x 轴、y 轴方向上的平移量
* 将平移矩阵与几何图形的每个顶点相乘，实现齐次平移变换
*******************************/
void my_traslate_homogeneous(struct my_v_inhomogeneous* polygon,
                    int polygon_vertex_count, int tx, int ty)
{
    //装配生成平移矩阵
    float translate_matrix[3][3];
    memset(translate_matrix, 0, sizeof(int) * 9);
    translate_matrix[0][0] = translate_matrix[1][1] = translate_matrix[2][2] = 1;
    translate_matrix[0][2] = tx;
```

```
        translate_matrix[1][2] = ty;

        //遍历并平移多边形的每个顶点
        for (int vIndex = 0; vIndex < polygon_vertex_count; vIndex++)
        {
            struct my_v_homogeneous input_v;
            input_v.x = polygon[vIndex].x;
            input_v.y = polygon[vIndex].y;
            input_v.ratio = 1;
            input_v = matrix_multiply_vector(translate_matrix, input_v);
            //平移矩阵作用到每个顶点,即矩阵顶点(向量)相乘
            polygon[vIndex].x = input_v.x;
            polygon[vIndex].y = input_v.y;
        }
    }
```

此处对 matrix_multiply_vector 函数做一个简单介绍,相应代码如下:

```
/***************************************
 * 矩阵向量相乘算法——4×4 矩阵与向量相乘算法类似
 * matrix[][3]:变换矩阵;input_v:几何图形的顶点数量
 * 将平移矩阵与几何图形的每个顶点相乘,实现齐次平移变换
 ****************************************/
struct my_v_homogeneous matrix_multiply_vector(float matrix[][3],
                                                struct my_v_homogeneous input_v)
{
    struct my_v_homogeneous translated_v;
    translated_v.x = matrix[0][0] * input_v.x + matrix[0][1] * input_v.y + matrix[0][2] * 1;
    translated_v.y = matrix[1][0] * input_v.x + matrix[1][1] * input_v.y + matrix[1][2] * 1;
    translated_v.ratio = matrix[2][0]*input_v.x +matrix[2][1]*input_v.y +matrix[2][2] * 1;
    return translated_v;
}
```

3. 案例效果

应用式(4.9)进行平移变换,对如图 4-8(a)所示的中心在原点的三角形,向 x 轴正方向平移 40 个单位,向 y 轴正方向平移 50 个单位,最终效果如图 4-8(b)所示。

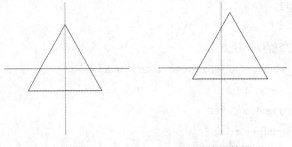

齐次平移变换

(a) 初始图形　　(b) 平移后的图形

图 4-8　齐次平移变换示意图

4.2.7　齐次缩放变换

二维图形齐次缩放的参数化表达式如下：

$$\begin{cases} x' = s_x \times x \\ y' = s_y \times y \end{cases} \tag{4.10}$$

其中，s_x、s_y 分别表示物体在 x、y 轴方向上的缩放量。

对应的齐次平移变换的矩阵乘积形式如下：

$$\begin{pmatrix} x' \\ y' \\ 1 \end{pmatrix} = \begin{pmatrix} s_x & 0 & 0 \\ 0 & s_y & 0 \\ 0 & 0 & 1 \end{pmatrix} \begin{pmatrix} x \\ y \\ 1 \end{pmatrix} \tag{4.11}$$

1. 变换函数代码实现

代码如下：

```
/*******************************************
* 齐次缩放变换
* polygon：几何图形；polygon_vertex_count：几何图形的顶点数量；
* sx、sy：x 轴、y 轴方向上的缩放系数
* 将缩放矩阵与几何图形的每个顶点相乘，实现齐次缩放变换
*******************************************/
void my_scale_homogeneous(struct my_v_inhomogeneous* polygon,
                          int polygon_vertex_count, float sx, float sy)
{
    //装配生成缩放矩阵
    float translate_matrix[3][3];
```

第 4 章 二维几何变换

```
memset(translate_matrix, 0, sizeof(int) * 9);
translate_matrix[0][0] = sx;
translate_matrix[1][1] = sy;
translate_matrix[2][2] = 1;

//遍历并缩放多边形的每个顶点
for (int vIndex = 0; vIndex < polygon_vertex_count; vIndex++)
{
    struct my_v_homogeneous input_v;
    input_v.x = polygon[vIndex].x;
    input_v.y = polygon[vIndex].y;
    input_v.ratio = 1;
    input_v = matrix_multiply_vector(translate_matrix, input_v);
    polygon[vIndex].x = input_v.x;
    polygon[vIndex].y = input_v.y;
}
}
```

2. 案例效果

应用式(4.11)进行缩放变换，对如图 4-9(a)所示的中心在原点的三角形，在 x 轴方向上缩放 1.5 倍，在 y 轴方向上缩放 0.5 倍，最终效果如图 4-9(b)所示。

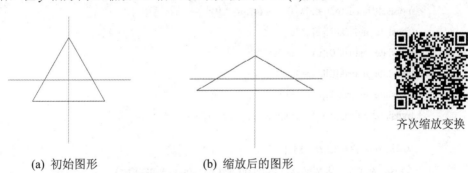

齐次缩放变换

(a) 初始图形　　　　(b) 缩放后的图形

图 4-9　齐次缩放变换示意图

4.2.8　齐次旋转变换

二维图形齐次旋转的参数化表达式如下：

$$\begin{cases} x' = x\cos\theta - y\sin\theta \\ y' = x\sin\theta + y\cos\theta \end{cases} \tag{4.12}$$

其中，θ表示物体绕点逆时针旋转的角度。对应的齐次旋转变换的矩阵乘积形式如下：

$$\begin{pmatrix} x' \\ y' \\ 1 \end{pmatrix} = \begin{pmatrix} \cos\theta & -\sin\theta & 0 \\ \sin\theta & \cos\theta & 0 \\ 0 & 0 & 1 \end{pmatrix} \begin{pmatrix} x \\ y \\ 1 \end{pmatrix} \tag{4.13}$$

1. 变换函数代码实现

代码如下：

```
/*************************************
 * 齐次旋转变换
 * polygon：几何图形；polygon_vertex_count：几何图形的顶点数量；
 * angle：几何图形的旋转角度，正值逆时针旋转，负值顺时针旋转
 * 将旋转矩阵与几何图形的每个顶点相乘，实现齐次旋转变换
 *************************************/
void my_rotate_homogeneous(struct my_v_inhomogeneous* polygon,
                    int polygon_vertex_count, float angle)
{
    angle = angle * 3.14159 / 180;
    //装配生成旋转矩阵
    float translate_matrix[3][3];
    memset(translate_matrix, 0, sizeof(int) * 9);
    translate_matrix[2][2] = 1;
    translate_matrix[0][0] = cos(angle);
    translate_matrix[0][1] = -sin(angle);
    translate_matrix[1][0] = sin(angle);
    translate_matrix[1][1] = cos(angle);

    //遍历并旋转多边形的每个顶点
    for (int vIndex = 0; vIndex < polygon_vertex_count; vIndex++)
    {
        struct my_v_homogeneous input_v;
        input_v.x = polygon[vIndex].x;
        input_v.y = polygon[vIndex].y;
        input_v.ratio = 1;
```

```
        input_v = matrix_multiply_vector(translate_matrix, input_v);
        polygon[vIndex].x = input_v.x;
        polygon[vIndex].y = input_v.y;
    }
}
```

2. 案例效果

应用式(4.13)进行旋转变换,对如图 4-10(a)所示的中心在原点的三角形逆时针旋转 50°,最终效果如图 4-10(b)所示。

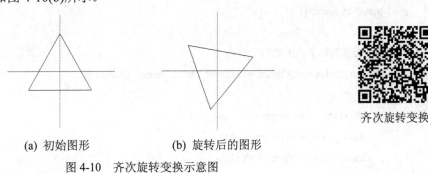

(a) 初始图形　　　　(b) 旋转后的图形

图 4-10　齐次旋转变换示意图

齐次旋转变换

4.2.9　齐次对称变换

1. 关于 x 轴齐次对称

二维图形关于 x 轴的齐次对称变换的参数化表达式如下:

$$\begin{cases} x' = x \\ y' = -y \end{cases} \tag{4.14}$$

对应的齐次对称变换的矩阵乘积形式如下:

$$\begin{pmatrix} x' \\ y' \\ 1 \end{pmatrix} = \begin{pmatrix} 1 & 0 & 0 \\ 0 & -1 & 0 \\ 0 & 0 & 1 \end{pmatrix} \begin{pmatrix} x \\ y \\ 1 \end{pmatrix} \tag{4.15}$$

1) 变换函数代码实现

代码如下:

```
/*******************************
 * 关于 x 轴的齐次对称变换
 * polygon: 几何图形; polygon_vertex_count: 几何图形的顶点数量
 * 将对称矩阵与几何图形的每个顶点相乘,实现齐次对称变换
```

***/
void my_reflection_x_homogeneous(struct my_v_inhomogeneous* polygon,
                    int polygon_vertex_count)
{
    //装配生成对称矩阵
    float translate_matrix[3][3];
    memset(translate_matrix, 0, sizeof(int) * 9);
    translate_matrix[0][0] = translate_matrix[2][2] = 1;
    translate_matrix[1][1] = -1;

    //遍历并对称多边形的每个顶点
    for (int vIndex = 0; vIndex < polygon_vertex_count; vIndex++)
    {
        struct my_v_homogeneous input_v;
        input_v.x = polygon[vIndex].x;
        input_v.y = polygon[vIndex].y;
        input_v.ratio = 1;
        input_v = matrix_multiply_vector(translate_matrix, input_v);
        polygon[vIndex].x = input_v.x;
        polygon[vIndex].y = input_v.y;
    }
}
```

2) 案例效果

应用式(4.15)进行关于 $x$ 轴的对称变换，对如图 4-11(a)所示的中心在原点的三角形关于 $x$ 轴进行对称变换，最终效果如图 4-11(b)所示。

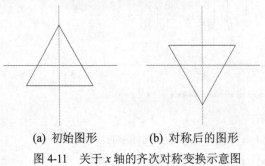
(a) 初始图形　　(b) 对称后的图形

图 4-11　关于 $x$ 轴的齐次对称变换示意图

关于 $x$ 轴的齐次对称
```

2. 关于 y 轴齐次对称

二维图形关于 y 轴的齐次对称变换的参数化表达式如下：

$$\begin{cases} x' = -x \\ y' = y \end{cases} \tag{4.16}$$

对应的齐次对称变换的矩阵乘积形式如下：

$$\begin{pmatrix} x' \\ y' \\ 1 \end{pmatrix} = \begin{pmatrix} -1 & 0 & 0 \\ 0 & 1 & 0 \\ 0 & 0 & 1 \end{pmatrix} \begin{pmatrix} x \\ y \\ 1 \end{pmatrix} \tag{4.17}$$

1) 变换函数代码实现

代码如下：

```
/*****************************************
 * 关于 y 轴的齐次对称变换
 * polygon：几何图形；polygon_vertex_count：几何图形的顶点数量
 * 将对称矩阵与几何图形的每个顶点相乘，实现齐次对称变换
 *****************************************/
void my_reflection_y_homogeneous(struct my_v_inhomogeneous* polygon,
                                 int polygon_vertex_count)
{
    //装配生成对称矩阵
    float translate_matrix[3][3];
    memset(translate_matrix, 0, sizeof(int) * 9);
    translate_matrix[1][1] = translate_matrix[2][2] = 1;
    translate_matrix[0][0] = -1;

    //遍历并对称多边形的每个顶点
    for (int vIndex = 0; vIndex < polygon_vertex_count; vIndex++)
    {
        struct my_v_homogeneous input_v;
        input_v.x = polygon[vIndex].x;
        input_v.y = polygon[vIndex].y;
        input_v.ratio = 1;
        input_v = matrix_multiply_vector(translate_matrix, input_v);
        polygon[vIndex].x = input_v.x;
```

```
        polygon[vIndex].y = input_v.y;
    }
}
```

2) 案例效果

应用式(4.17)进行关于 y 轴的对称变换，对如图 4-12(a)所示的中心在原点的三角形关于 y 轴进行对称变换，最终效果如图 4-12(b)所示。

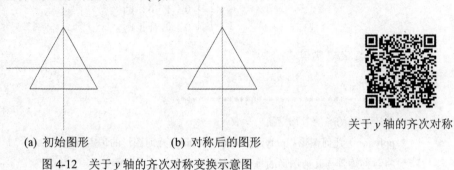

关于 y 轴的齐次对称

(a) 初始图形　　　　(b) 对称后的图形

图 4-12　关于 y 轴的齐次对称变换示意图

4.2.10　齐次错切变换

1. 相对于 x 轴方向上的齐次错切

二维图形关于 x 轴的齐次错切变换的参数化表达式如下：

$$\begin{cases} x' = x + \text{sh}_x \times y \\ y' = y \end{cases} \tag{4.18}$$

其中，sh_x 为错切参数。

对应的齐次错切变换的矩阵乘积形式如下：

$$\begin{pmatrix} x' \\ y' \\ 1 \end{pmatrix} = \begin{pmatrix} 1 & \text{sh}_x & 0 \\ 0 & 1 & 0 \\ 0 & 0 & 1 \end{pmatrix} \begin{pmatrix} x \\ y \\ 1 \end{pmatrix} \tag{4.19}$$

1) 变换函数代码实现

代码如下：

```
/*******************************
 * 相对于 x 轴方向的齐次错切
 * polygon：几何图形；polygon_vertex_count：几何图形的顶点数量；shx：切变程度
 * 将错切矩阵与几何图形的每个顶点相乘，实现齐次错切变换
 *******************************/
```

```
void my_shear_x_homogeneous(struct my_v_inhomogeneous* polygon,
                    int polygon_vertex_count, float shx)
{
    //装配生成错切矩阵
    float translate_matrix[3][3];
    memset(translate_matrix, 0, sizeof(int) * 9);
    translate_matrix[0][0] = translate_matrix[1][1] = translate_matrix[2][2] = 1;
    translate_matrix[0][1] = shx;

    //遍历并错切多边形的每个顶点
    for (int vIndex = 0; vIndex < polygon_vertex_count; vIndex++)
    {
        struct my_v_homogeneous input_v;
        input_v.x = polygon[vIndex].x;
        input_v.y = polygon[vIndex].y;
        input_v.ratio = 1;
        input_v = matrix_multiply_vector(translate_matrix, input_v);
        polygon[vIndex].x = input_v.x;
        polygon[vIndex].y = input_v.y;
    }
}
```

2) 案例效果

应用式(4.19)进行齐次错切变换，对如图 4-13(a)所示的中心在原点的三角形，设置 x 轴方向的齐次错切参数为 1.2，最终效果如图 4-13(b)所示。

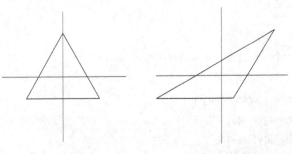

(a) 初始图形　　　　　(b) 错切后的图形　　　　　在 x 轴方向上的齐次错切

图 4-13　在 x 轴方向上的齐次错切变换示意图

2. 相对于 y 轴方向上的齐次错切

二维图形关于 y 轴的齐次错切变换的参数化表达式如下：

$$\begin{cases} x' = x \\ y' = sh_y \times x + y \end{cases} \quad (4.20)$$

其中，sh_y 为错切参数。

对应的齐次错切变换的矩阵乘积形式如下：

$$\begin{pmatrix} x' \\ y' \\ 1 \end{pmatrix} = \begin{pmatrix} 1 & 0 & 0 \\ sh_y & 1 & 0 \\ 0 & 0 & 1 \end{pmatrix} \begin{pmatrix} x \\ y \\ 1 \end{pmatrix} \quad (4.21)$$

1) 变换函数代码实现

代码如下：

```
/*******************************
 * 相对于 y 轴方向的齐次错切(变换 y 坐标的错切)
 * polygon：几何图形；polygon_vertex_count：几何图形的顶点数量；shy：切变程度
 * 将错切矩阵与几何图形的每个顶点相乘，实现齐次错切变换
 *******************************/
void my_shear_y_homogeneous(struct my_v_inhomogeneous* polygon,
                            int polygon_vertex_count, float shy)
{
    //装配生成错切矩阵
    float translate_matrix[3][3];
    memset(translate_matrix, 0, sizeof(int) * 9);
    translate_matrix[0][0] = translate_matrix[1][1] = translate_matrix[2][2] = 1;
    translate_matrix[1][0] = shy;

    //遍历并错切多边形的每个顶点
    for (int vIndex = 0; vIndex < polygon_vertex_count; vIndex++)
    {
        struct my_v_homogeneous input_v;
        input_v.x = polygon[vIndex].x;
        input_v.y = polygon[vIndex].y;
        input_v.ratio = 1;
        input_v = matrix_multiply_vector(translate_matrix, input_v);
```

```
        polygon[vIndex].x = input_v.x;
        polygon[vIndex].y = input_v.y;
    }
}
```

2) 案例效果

应用式(4.21)进行齐次错切变换，对如图 4-14(a)所示的中心在原点的三角形，设置 y 轴方向的齐次错切参数为 0.8，最终效果如图 4-14(b)所示。

(a) 初始图形　　(b) 错切后的图形

图 4-14　在 y 轴方向上的齐次错切变换示意图

在 y 轴方向上的齐次错切

4.3　二维复合几何变换

4.3.1　绕任意点旋转变换

二维图形绕任意点旋转的效果等价于多个二维基本几何变换的有序复合。此处，假定要实现任一二维图形绕点 $P(x, y)$ 逆时针旋转 θ，其复合变换公式如下：

$$\begin{pmatrix} x' \\ y' \\ 1 \end{pmatrix} = \boldsymbol{CBA} \begin{pmatrix} x \\ y \\ 1 \end{pmatrix} \tag{4.22}$$

其中：$\boldsymbol{A}$ 为平移矩阵，对应式(4.9)，其达到的效果为二维图形在 x 轴、y 轴方向上平移 $(-x, -y)$，使得 P 点与原点重合；$\boldsymbol{B}$ 为旋转矩阵，对应式(4.13)，其达到的效果为平移后的二维图形绕原点逆时针旋转 θ；$\boldsymbol{C}$ 为平移矩阵 $\boldsymbol{A}$ 的逆变换，对应式(4.9)，即将旋转后的二维图形在 x 轴、y 轴方向上平移 (x, y)。

1. 变换函数代码实现

代码如下：

```
/*********************************
```

```
* 绕任意点旋转变换
* polygon：几何图形；polygon_vertex_count：几何图形的顶点数量；x、y：任意点的坐标；
* angle：几何图形的旋转角度
******************************************/
void my_rotateArbitraryAxis_homogeneous(struct my_v_inhomogeneous* polygon, int
                              polygon_vertex_count, double x, double y, float angle)
{
    //计算平移量
    double tx = -x;
    double ty = -y;

    my_traslate_homogeneous(polygon, polygon_vertex_count, tx, ty);         //平移变换
    my_rorate_homogeneous(polygon, polygon_vertex_count, angle);            //旋转变换
    my_traslate_homogeneous(polygon, polygon_vertex_count, -tx, -ty);       //平移回原位
}
```

2. 案例效果

采用绕任意点旋转变换，对如图 4-15(a)所示的中心在原点的三角形绕点 P(−20，30)逆时针旋转 30°，最终效果如图 4-15(b)所示。

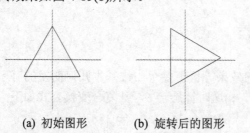

(a) 初始图形　　(b) 旋转后的图形

图 4-15　采用绕任意点旋转变换示意图

绕任意点旋转变换

4.3.2　相对任意点缩放变换

二维图形相对任意点缩放的效果等价于多个二维基本几何变换的有序复合。此处，假定二维图形相对点 P(x, y)进行缩放，其复合变换公式如下：

$$\begin{pmatrix} x' \\ y' \\ 1 \end{pmatrix} = CBA \begin{pmatrix} x \\ y \\ 1 \end{pmatrix} \tag{4.23}$$

其中：A 为平移矩阵，对应式(4.9)，使二维图形及 P 在 x 轴、y 轴方向上平移$(-x, -y)$，达到 P 点与原点重合；B 为缩放矩阵，对应式(4.11)，使二维图形绕原点缩放；C 为平移矩阵 A 的逆变换，对应式(4.9)，使得二维图形及 P 在 x 轴、y 轴方向上平移(x, y)，达到 P 点与回到原来的位置。

1. 变换函数代码实现

代码如下：

```
/********************************
 * 相对任意点缩放变换
 * polygon：几何图形；polygon_vertex_count：几何图形的顶点数量；x、y：任意点的坐标；
 * sx、sy：x 轴、y 轴方向上的缩放量
 ********************************/
void my_scaleArbitraryPoint_homogeneous(struct my_v_inhomogeneous* polygon,
                    int polygon_vertex_count, double x, double y,float sx, float sy )
{
    //计算平移量
    float tx = -x;
    float ty = -y;
    my_traslate_homogeneous(polygon, polygon_vertex_count, tx, ty);        //平移变换
    my_scale_homogeneous(polygon, polygon_vertex_count, sx, sy);           //比例缩放
    my_traslate_homogeneous(polygon, polygon_vertex_count, -tx, -ty);      //反向平移
}
```

2. 案例效果

采用相对任意点缩放变换，对如图 4-16(a)所示的中心在原点的三角形关于点$(-100, -100)$进行缩放，x、y 轴上的缩放系数分别为 1.2、0.5，最终效果如图 4-16(b)所示。

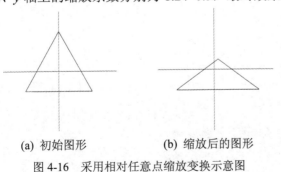

(a) 初始图形　　　　(b) 缩放后的图形

相对任意点缩放变换

图 4-16　采用相对任意点缩放变换示意图

4.4 课外拓展性实验

将如下复合变换分解为两个基本几何变换，对三角形分别实施这两个几何变换，观察三角形的变化，并讨论能否交换几何变换的次序。

$$\begin{pmatrix} 1/2 & \sqrt{3}/2 & 10 \\ -\sqrt{3}/2 & 1/2 & 5 \\ 0 & 0 & 1 \end{pmatrix}$$

课外拓展性实验

第 5 章 三维几何变换

5.1 实验内容简述和实验目标

基本实验内容包括：三维基本几何变换(三维平移变换、三维旋转变换、三维缩放变换、三维错切变换、三维对称变换)、三维复合几何变换(三维图形绕空间任意轴旋转、三维图形相对任意点缩放)和三维几何变换综合示例。同时，配备了一个课外拓展性实验。

完成本实验后，读者能够：

(1) 推导和熟记齐次坐标系下三维图形几何变换对应的代数表示方式(布鲁姆知识模型：记忆和理解)；

(2) 推导和写出三维图形绕空间任意轴旋转的变换矩阵(布鲁姆知识模型：理解和应用)；

(3) 判断给定矩阵复合对三维图形产生的几何变换(布鲁姆知识模型：应用、分析和评价)；

(4) 发现并排除三维图形几何变换后三维图形无法在窗口显示的原因——与视点和投影变换紧密关联(布鲁姆知识模型：应用和分析)；

(5) 结合 OpenGL 编程实现——通过鼠标、键盘交互，展现给定三维图形的基本几何变换、任意复合几何变换、视点变换和投影变换(布鲁姆知识模型：应用)。

5.2 三维基本几何变换

5.2.1 三维平移变换

三维图形平移变换的参数化表达式如下：

$$\begin{cases} x' = x + t_x \\ y' = y + t_y \\ z' = z + t_z \end{cases} \tag{5.1}$$

其中，t_x、t_y、t_z 分别表示三维图形在 x、y、z 轴方向上的平移量。

对应的齐次平移变换的矩阵乘积形式如下：

$$\begin{pmatrix} x' \\ y' \\ z' \\ 1 \end{pmatrix} = \begin{pmatrix} 1 & 0 & 0 & t_x \\ 0 & 1 & 0 & t_y \\ 0 & 0 & 1 & t_z \\ 0 & 0 & 0 & 1 \end{pmatrix} \begin{pmatrix} x \\ y \\ z \\ 1 \end{pmatrix} \tag{5.2}$$

1. 关键数据结构

自定义如下数据结构用以表示三维齐次坐标下的顶点和向量。该数据结构应用于本章所有齐次三维几何变换实验。

```
struct my_v_homogeneous
{
    float x;
    float y;
    float z;
    float ratio;
};
```

2. 变换函数代码实现

代码如下：

```
/*****************************
 * 齐次平移变换
 * polygon：几何图形；vertex_count：几何图形的顶点数量；
 * tx、ty、tz：x 轴、y 轴、z 轴方向上的平移量
 * 将平移矩阵与几何图形的每个顶点相乘，实现齐次平移变换
 *****************************/
void my_translate_homogeneous(struct my_v_homogeneous* polygon, int vertex_count,
                    float tx, float ty, float tz)
{
    //将式(5.2)装配生成平移矩阵
    float translate_matrix[4][4];
    memset(translate_matrix, 0, sizeof(int) * 16);
    translate_matrix[0][0] = translate_matrix[1][1] = translate_matrix[2][2] =
    translate_matrix[3][3] = 1;         //对角线赋值为 1
```

```
translate_matrix[0][3] = tx;
translate_matrix[1][3] = ty;
translate_matrix[2][3] = tz;

//遍历三维图形的每个顶点并平移每个顶点
for (int vIndex = 0; vIndex < vertex_count; vIndex++)
{
    struct my_v_homogeneous input_v;
    input_v.x = polygon[vIndex].x;
    input_v.y = polygon[vIndex].y;
    input_v.z = polygon[vIndex].z;
    input_v.ratio = 1;
    polygon[vIndex] = matrix_multiply_vector(translate_matrix, polygon[vIndex]);
    polygon[vIndex].x = input_v.x;
    polygon[vIndex].y = input_v.y;
    polygon[vIndex].z = input_v.z;
    polygon[vIndex].ratio = input_v.ratio;
}
```

3. 案例效果

应用式(5.2)对如图 5-1(a)所示的顶点在原点的长方体向 x 轴正方向移动 30 个单位，最终效果如图 5-1(b)所示。

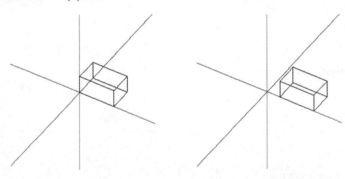

(a) 初始图形　　　　　(b) 平移后的图形

图 5-1　三维平移变换示意图

三维平移变换

5.2.2 三维旋转变换

1. 绕 x 轴旋转

三维空间中的任意三维图形，绕 x 轴的逆时针方向旋转 θ，对应的参数化表达式如下：

$$\begin{cases} x' = x \\ y' = y \times \cos\theta - z \times \sin\theta \\ z' = y \times \sin\theta + z \times \cos\theta \end{cases} \tag{5.3}$$

对应的齐次旋转变换的矩阵乘积形式如下：

$$\begin{pmatrix} x' \\ y' \\ z' \\ 1 \end{pmatrix} = \begin{pmatrix} 1 & 0 & 0 & 0 \\ 0 & \cos\theta & -\sin\theta & 0 \\ 0 & \sin\theta & \cos\theta & 0 \\ 0 & 0 & 0 & 1 \end{pmatrix} \begin{pmatrix} x \\ y \\ z \\ 1 \end{pmatrix} = R(\theta) \begin{pmatrix} x \\ y \\ z \\ 1 \end{pmatrix} \tag{5.4}$$

1) 变换函数代码实现

代码如下：

```
/*******************************
 * 绕 x 轴的齐次旋转变换
 * polygon：几何图形；vertex_count：几何图形的顶点数量；angle：几何图形的旋转角度
 * 将旋转矩阵与几何图形的每个顶点相乘，实现绕 x 轴的齐次旋转变换
 *******************************/
void my_rotate_x_homogeneous(struct my_v_homogeneous* polygon, int vertex_count,
                             float angle)
{
    //将式(5.4)装配生成旋转矩阵
    float translate_matrix[4][4];
    memset(translate_matrix, 0, sizeof(int) * 16);
    float pai = 3.14159;

    angle = angle * pai / 180;
    translate_matrix[0][0] = 1;
    translate_matrix[1][1] = cos(angle);
    translate_matrix[1][2] = -sin(angle);
    translate_matrix[2][1] = sin(angle);
```

```
translate_matrix[2][2] = cos(angle);
translate_matrix[3][3] = 1;

//遍历三维图形的每个顶点并旋转每个顶点
for (int vIndex = 0; vIndex < vertex_count; vIndex++)
{
    struct my_v_homogeneous input_v;
    input_v.x = polygon[vIndex].x;
    input_v.y = polygon[vIndex].y;
    input_v.z = polygon[vIndex].z;
    input_v.ratio = 1;
    input_v = matrix_multiply_vector(translate_matrix, input_v);
    polygon[vIndex].x = input_v.x;
    polygon[vIndex].y = input_v.y;
    polygon[vIndex].z = input_v.z;
    polygon[vIndex].ratio = input_v.ratio;
}
}
```

2) 案例效果

应用式(5.4)对如图 5-2(a)所示的顶点在原点的长方体绕 x 轴逆时针旋转 60°，最终效果如图 5-2(b)所示。

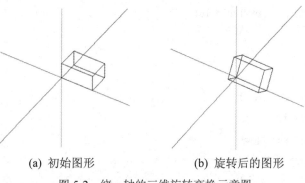

(a) 初始图形　　(b) 旋转后的图形　　三维旋转变换
(绕 x 轴旋转)

图 5-2　绕 x 轴的三维旋转变换示意图

2. 绕 y 轴旋转

三维空间中的任意三维图形，绕 y 轴的逆时针方向旋转 θ，对应的参数化表达式如下：

$$\begin{cases} x' = x \times \cos\theta + z \times \sin\theta \\ y' = y \\ z' = -x \times \sin\theta + z \times \cos\theta \end{cases} \tag{5.5}$$

对应的齐次旋转变换的矩阵乘积形式如下：

$$\begin{pmatrix} x' \\ y' \\ z' \\ 1 \end{pmatrix} = \begin{pmatrix} \cos\theta & 0 & \sin\theta & 0 \\ 0 & 1 & 0 & 0 \\ -\sin\theta & 0 & \cos\theta & 0 \\ 0 & 0 & 0 & 1 \end{pmatrix} \begin{pmatrix} x \\ y \\ z \\ 1 \end{pmatrix} = R(\theta) \begin{pmatrix} x \\ y \\ z \\ 1 \end{pmatrix} \tag{5.6}$$

1) 变换函数代码实现

代码如下：

```
/*************************************
* 绕 y 轴的齐次旋转变换
* polygon：几何图形；vertex_count：几何图形的顶点数量；angle：几何图形的旋转角度
* 将旋转矩阵与几何图形的每个顶点相乘，实现绕 y 轴的齐次旋转变换
*************************************/
void my_rotate_y_homogeneous(struct my_v_homogeneous* polygon, int vertex_count,
                float angle)
{
    //将式(5.6)装配生成旋转矩阵
    float translate_matrix[4][4];
    memset(translate_matrix, 0, sizeof(int) * 16);
    float pai = 3.14159;

    angle = angle * pai / 180;
    translate_matrix[0][0] = cos(angle);
    translate_matrix[1][1] = 1;
    translate_matrix[0][2] = sin(angle);
    translate_matrix[2][0] = -sin(angle);
    translate_matrix[2][2] = cos(angle);
    translate_matrix[3][3] = 1;

    //遍历三维图形的每个顶点并旋转每个顶点
```

```
for (int vIndex = 0; vIndex < vertex_count; vIndex++)
{
    struct my_v_homogeneous input_v;
    input_v.x = polygon[vIndex].x;
    input_v.y = polygon[vIndex].y;
    input_v.z = polygon[vIndex].z;
    input_v.ratio = 1;
    input_v = matrix_multiply_vector(translate_matrix, input_v);
    polygon[vIndex].x = input_v.x;
    polygon[vIndex].y = input_v.y;
    polygon[vIndex].z = input_v.z;
    polygon[vIndex].ratio = input_v.ratio;
}
}
```

2) 案例效果

应用式(5.6)对如图 5-3(a)所示的顶点在原点的长方体绕 y 轴逆时针旋转 $60°$,最终效果如图 5-3(b)所示。

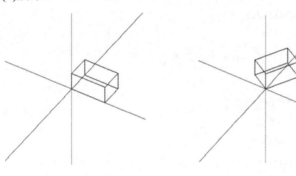

  (a) 初始图形   (b) 旋转后的图形

三维旋转变换
(绕 y 轴旋转)

图 5-3 绕 y 轴的三维旋转变换示意图

3. 绕 z 轴旋转

三维图形绕 z 轴逆时针方向旋转 θ,对应的参数化表达式如下:

$$\begin{cases} x' = x \times \cos\theta - y \times \sin\theta \\ y' = x \times \sin\theta + y \times \cos\theta \\ z' = z \end{cases} \tag{5.7}$$

对应的齐次旋转变换的矩阵乘积形式如下：

$$\begin{pmatrix} x' \\ y' \\ z' \\ 1 \end{pmatrix} = \begin{pmatrix} \cos\theta & -\sin\theta & 0 & 0 \\ \sin\theta & \cos\theta & 0 & 0 \\ 0 & 0 & 1 & 0 \\ 0 & 0 & 0 & 1 \end{pmatrix} \begin{pmatrix} x \\ y \\ z \\ 1 \end{pmatrix} \tag{5.8}$$

1) 变换函数代码实现

代码如下：

```
/*****************************************
* 绕 z 轴的齐次旋转变换
* polygon：几何图形；vertex_count：几何图形的顶点数量；angle：几何图形的旋转角度
* 将旋转矩阵与几何图形的每个顶点相乘，实现绕 z 轴的齐次旋转变换
*****************************************/
void my_rotate_z_homogeneous(struct my_v_homogeneous* polygon, int vertex_count,
                 float angle)
{
    //将式(5.8)装配生成旋转矩阵
    float translate_matrix[4][4];
    memset(translate_matrix, 0, sizeof(int) * 16);
    float pai = 3.14159;

    angle = angle * pai / 180;
    translate_matrix[0][0] = cos(angle);
    translate_matrix[1][1] = cos(angle);
    translate_matrix[0][1] = -sin(angle);
    translate_matrix[1][0] = sin(angle);
    translate_matrix[2][2] = 1;
    translate_matrix[3][3] = 1;

    //遍历三维图形的每个顶点并旋转每个顶点
    for (int vIndex = 0; vIndex < vertex_count; vIndex++)
    {
        struct my_v_homogeneous input_v;
        input_v.x = polygon[vIndex].x;
```

```
            input_v.y = polygon[vIndex].y;
            input_v.z = polygon[vIndex].z;
            input_v.ratio = 1;
            input_v = matrix_multiply_vector(translate_matrix, input_v);
            polygon[vIndex].x = input_v.x;
            polygon[vIndex].y = input_v.y;
            polygon[vIndex].z = input_v.z;
            polygon[vIndex].ratio = input_v.ratio;
        }
    }
```

2) 案例效果

应用式(5.8)对如图 5-4(a)所示的顶点在原点的长方体绕 z 轴逆时针旋转 $60°$，最终效果如图 5-4(b)所示。

(a) 初始图形　　　　(b) 旋转后的图形

三维旋转变换
(绕 z 轴旋转)

图 5-4　绕 z 轴的三维旋转变换示意图

5.2.3　三维缩放变换

三维图形相对原点缩放的参数化表达式如下：

$$\begin{cases} x' = s_x \times x \\ y' = s_y \times y \\ z' = s_z \times z \end{cases} \tag{5.9}$$

其中，s_x、s_y、s_z 分别表示三维图形在 x 轴、y 轴和 z 轴上的缩放系数。

对应的齐次缩放变换的矩阵乘积形式如下：

$$\begin{pmatrix} x' \\ y' \\ z' \\ 1 \end{pmatrix} = \begin{pmatrix} s_x & 0 & 0 & 0 \\ 0 & s_y & 0 & 0 \\ 0 & 0 & s_z & 0 \\ 0 & 0 & 0 & 1 \end{pmatrix} \begin{pmatrix} x \\ y \\ z \\ 1 \end{pmatrix} \tag{5.10}$$

1. 变换函数代码实现

代码如下：

```
/*****************************************
 * 齐次缩放变换
 * polygon：几何图形；vertex_count：几何图形的顶点数量；
 * sx、sy、sz：x 轴、y 轴、z 轴方向上的缩放系数
 * 将缩放矩阵与几何图形的每个顶点相乘，实现齐次缩放变换
 *****************************************/
void my_scale_homogeneous(struct my_v_homogeneous* polygon, int vertex_count,
                          float sx, float sy, float sz)
{
    //将式(5.10)装配生成缩放矩阵
    float translate_matrix[4][4];
    memset(translate_matrix, 0, sizeof(int) * 16);
    translate_matrix[0][0] = sx;
    translate_matrix[1][1] = sy;
    translate_matrix[2][2] = sz;
    translate_matrix[3][3] = 1;

    //遍历三维图形的每个顶点并缩放多边形的每个顶点
    for (int vIndex = 0; vIndex < vertex_count; vIndex++)
    {
        struct my_v_homogeneous input_v;
        input_v.x = polygon[vIndex].x;
        input_v.y = polygon[vIndex].y;
        input_v.z = polygon[vIndex].z;
        input_v.ratio = 1;
        input_v = matrix_multiply_vector(translate_matrix, input_v);
        polygon[vIndex].x = input_v.x;
        polygon[vIndex].y = input_v.y;
        polygon[vIndex].z = input_v.z;
        polygon[vIndex].ratio = input_v.ratio;
    }
```

}

2. 案例效果

应用式(5.10)对如图 5-5(a)所示的一个顶点在原点的长方体进行错切变换,设置其在 x、y、z 轴方向上分别缩放 1.5 倍、1.2 倍、1.5 倍,最终效果如图 5-5(b)所示。

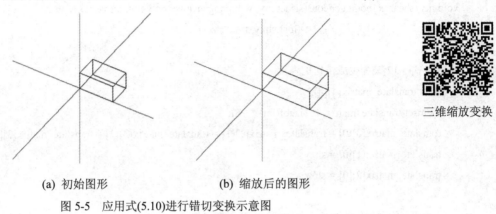

三维缩放变换

(a) 初始图形　　　　　(b) 缩放后的图形

图 5-5　应用式(5.10)进行错切变换示意图

5.2.4　三维错切变换

1. 依赖于 x 轴的齐次错切变换

三维图形以 x 轴为依赖轴的错切变换的参数化表达式如下:

$$\begin{cases} x' = x \\ y' = y + \text{sh}_{xy} \times x \\ z' = z + \text{sh}_{xz} \times x \end{cases} \tag{5.11}$$

保持三维图形上各顶点 x 坐标不变,y、z 坐标依 x 坐标,分别以 sh_{xy}、sh_{xz} 切变程度呈线性变换。

对应的齐次错切变换的矩阵乘积形式如下:

$$\begin{pmatrix} x' \\ y' \\ z' \\ 1 \end{pmatrix} = \begin{pmatrix} 1 & 0 & 0 & 0 \\ \text{sh}_{xy} & 1 & 0 & 0 \\ \text{sh}_{xz} & 0 & 1 & 0 \\ 0 & 0 & 0 & 1 \end{pmatrix} \begin{pmatrix} x \\ y \\ z \\ 1 \end{pmatrix} \tag{5.12}$$

1) 变换函数代码实现

代码如下:

/***

```
 * 依赖于 x 轴的齐次错切变换
 * polygon：几何图形；vertex_count：几何图形的顶点数量；shxy、shxz：相关切变程度
 * 将错切矩阵与几何图形的每个顶点相乘，实现依赖于 x 轴的齐次错切变换
**********************************/
void my_shear_x_homogeneous(struct my_v_homogeneous* polygon, int vertex_count,
                            float shxy, float shxz)
{
    //将式(5.12)装配生成错切矩阵
    float translate_matrix[4][4];
    memset(translate_matrix, 0, sizeof(int) * 16);
    translate_matrix[0][0] = translate_matrix[1][1] = translate_matrix[2][2] = translate_matrix[3][3] =1;
    translate_matrix[1][0] = shxy;
    translate_matrix[2][0] = shxz;

    //遍历三维图形的每个顶点并错切多边形的每个顶点
    for (int vIndex = 0; vIndex < vertex_count; vIndex++)
    {
        struct my_v_homogeneous input_v;
        input_v.x = polygon[vIndex].x;
        input_v.y = polygon[vIndex].y;
        input_v.z = polygon[vIndex].z;
        input_v.ratio = 1;
        input_v = matrix_multiply_vector(translate_matrix, input_v);
        polygon[vIndex].x = input_v.x;
        polygon[vIndex].y = input_v.y;
        polygon[vIndex].z = input_v.z;
        vpolygon[vIndex].ratio = input_v.ratio;
    }
}
```

2) 案例效果

应用式(5.12)对如图 5-6(a)所示的一个顶点在原点的长方体进行错切变换，设置其在 y、z 轴方向的切变程度分别为 0.3、0.6，最终效果如图 5-6(b)所示。

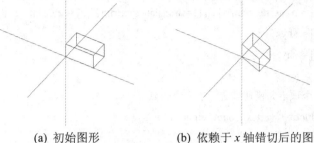

依赖于 x 轴的
齐次错切变换

(a) 初始图形　　　　(b) 依赖于 x 轴错切后的图形

图 5-6　依赖于 x 轴的齐次错切变换示意图

2. 依赖于 y 轴的齐次错切变换

三维图形以 y 轴为依赖轴的错切变换的参数化表达式如下：

$$\begin{cases} x' = x + \mathrm{sh}_{yx} \times y \\ y' = y \\ z' = z + \mathrm{sh}_{yz} \times y \end{cases} \tag{5.13}$$

保持三维图形上各顶点 y 坐标不变，x、z 坐标依 y 坐标分别以 sh_{yx}、sh_{yz} 切变程度呈线性变换。

对应的齐次错切变换的矩阵乘积形式如下：

$$\begin{pmatrix} x' \\ y' \\ z' \\ 1 \end{pmatrix} = \begin{pmatrix} 1 & \mathrm{sh}_{yx} & 0 & 0 \\ 0 & 1 & 0 & 0 \\ 0 & \mathrm{sh}_{yz} & 1 & 0 \\ 0 & 0 & 0 & 1 \end{pmatrix} \begin{pmatrix} x \\ y \\ z \\ 1 \end{pmatrix} \tag{5.14}$$

1) 变换函数代码实现

代码如下：

```
/*******************************
 * 依赖于 y 轴的齐次错切变换
 * polygon：几何图形；vertex_count：几何图形的顶点数量；shyx、shyz：相关切变程度
 * 将错切矩阵与几何图形的每个顶点相乘，实现依赖于 y 轴的齐次错切变换
 *******************************/
void my_shear_y_homogeneous(struct my_v_homogeneous* polygon, int vertex_count,
                            float shyx, float shyz)
{
    //将式(5.14)装配生成错切矩阵
    float translate_matrix[4][4];
    memset(translate_matrix, 0, sizeof(int) * 16);
```

```
translate_matrix[0][0] = translate_matrix[1][1] = translate_matrix[2][2] = translate_matrix[3][3] =1;
translate_matrix[0][1] = shyx;
translate_matrix[2][1] = shyz;

//遍历三维图形的每个顶点并错切多边形的每个顶点
for (int vIndex = 0; vIndex < vertex_count; vIndex++)
{
    struct my_v_homogeneous input_v;
    input_v.x = polygon[vIndex].x;
    input_v.y = polygon[vIndex].y;
    input_v.z = polygon[vIndex].z;
    input_v.ratio = 1;
    input_v = matrix_multiply_vector(translate_matrix, input_v);
    polygon[vIndex].x = input_v.x;
    polygon[vIndex].y = input_v.y;
    polygon[vIndex].z = input_v.z;
    polygon[vIndex].ratio = input_v.ratio;
}
```

2) 案例效果

应用式(5.14)对如图 5-7(a)所示的一个顶点在原点的长方体进行错切变换，设置其在 x、z 轴方向的切变程度分别为 0.4、1.2，最终效果如图 5-7(b)所示。

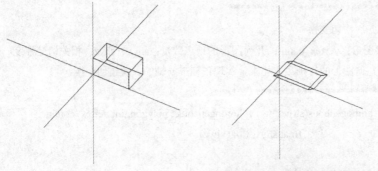

依赖于 y 轴的
齐次错切变换

(a) 初始图形　　　　(b) 依赖于 y 轴错切后的图形

图 5-7　依赖于 y 轴的齐次错切变换示意图

3. 依赖于 z 轴的齐次错切变换

三维图形以 z 轴为依赖轴的错切变换的参数化表达式如下：

$$\begin{cases} x' = x + \mathrm{sh}_{zx} \times z \\ y' = y + \mathrm{sh}_{zy} \times z \\ z' = z \end{cases} \quad (5.15)$$

保持三维图形上各顶点 z 坐标不变，x、y 坐标依 z 坐标分别以 sh_{zx}、sh_{zy} 切变程度呈线性变换。

对应的齐次错切变换的矩阵乘积形式如下：

$$\begin{pmatrix} x' \\ y' \\ z' \\ 1 \end{pmatrix} = \begin{pmatrix} 1 & 0 & \mathrm{sh}_{zx} & 0 \\ 0 & 1 & \mathrm{sh}_{zy} & 0 \\ 0 & 0 & 1 & 0 \\ 0 & 0 & 0 & 1 \end{pmatrix} \begin{pmatrix} x \\ y \\ z \\ 1 \end{pmatrix} \quad (5.16)$$

1) 变换函数代码实现

代码如下：

```
/************************************
* 依赖于 z 轴的齐次错切变换
* polygon：几何图形；vertex_count：几何图形的顶点数量；shzx、shzy：相关切变程度
* 将错切矩阵与几何图形的每个顶点相乘，实现依赖于 z 轴的齐次错切变换
************************************/
void my_shear_z_homogeneous(struct my_v_homogeneous* polygon, int vertex_count,
                float shzx, float shzy)
{
    //将式(5.16)装配生成错切矩阵
    float translate_matrix[4][4];
    memset(translate_matrix, 0, sizeof(int) * 16);
    translate_matrix[0][0] = translate_matrix[1][1] = translate_matrix[2][2] = translate_matrix[3][3] =1;
    translate_matrix[0][2] = shzx;
    translate_matrix[1][2] = shzy;

    //遍历三维图形的每个顶点并错切多边形的每个顶点
    for (int vIndex = 0; vIndex < vertex_count; vIndex++)
    {
        struct my_v_homogeneous input_v;
```

```
input_v.x = polygon[vIndex].x;
input_v.y = polygon[vIndex].y;
input_v.z = polygon[vIndex].z;
input_v.ratio = 1;
input_v = matrix_multiply_vector(translate_matrix, input_v);
polygon[vIndex].x = input_v.x;
polygon[vIndex].y = input_v.y;
polygon[vIndex].z = input_v.z;
polygon[vIndex].ratio = input_v.ratio;
    }
}
```

2) 案例效果

应用式(5.16)对如图 5-8(a)所示的一个顶点在原点的长方体进行错切变换，设置其在 x、z 轴方向的切变程度分别为 1.2、1.5，最终效果如图 5-8(b)所示。

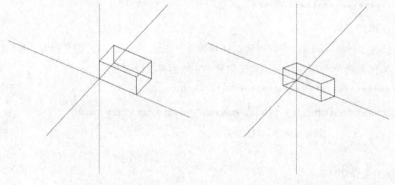

(a) 初始图形　　　　　　(b) 依赖于 z 轴错切后的图形

图 5-8　依赖于 z 轴的齐次错切变换示意图

依赖于 z 轴的齐次错切变换

5.2.5　三维对称变换

1. 关于 x 轴的齐次对称变换

三维图形关于 x 轴做对称变换的参数化表达式如下：

$$\begin{cases} x' = x \\ y' = -y \\ z' = -z \end{cases} \tag{5.17}$$

第 5 章　三维几何变换

对应的齐次对称变换的矩阵乘积形式如下：

$$\begin{pmatrix} x' \\ y' \\ z' \\ 1 \end{pmatrix} = \begin{pmatrix} 1 & 0 & 0 & 0 \\ 0 & -1 & 0 & 0 \\ 0 & 0 & -1 & 0 \\ 0 & 0 & 0 & 1 \end{pmatrix} \begin{pmatrix} x \\ y \\ z \\ 1 \end{pmatrix} \qquad (5.18)$$

1) 变换函数代码实现

代码如下：

```
/*****************************************
* 关于 x 轴的齐次对称变换
* polygon：几何图形；vertex_count：几何图形的顶点数量
* 将对称矩阵与几何图形的每个顶点相乘，实现关于 x 轴的齐次对称变换
*****************************************/
void my_reflection_x_homogeneous(struct my_v_homogeneous* polygon,
                                 int vertex_count)
{
    //将式(5.18)装配生成对称矩阵
    float translate_matrix[4][4];
    memset(translate_matrix, 0, sizeof(int) * 16);
    translate_matrix[0][0] = 1;
    translate_matrix[1][1] = -1;
    translate_matrix[2][2] = -1;
    translate_matrix[3][3] = 1;

    //遍历三维图形的每个顶点并对称多边形的每个顶点
    for (int vIndex = 0; vIndex < vertex_count; vIndex++)
    {
        struct my_v_homogeneous input_v;
        input_v.x = polygon[vIndex].x;
        input_v.y = polygon[vIndex].y;
        input_v.z = polygon[vIndex].z;
        input_v.ratio = 1;
        input_v = matrix_multiply_vector(translate_matrix, input_v);
        polygon[vIndex].x = input_v.x;
```

```
            polygon[vIndex].y = input_v.y;
            polygon[vIndex].z = input_v.z;
            polygon[vIndex].ratio = input_v.ratio;
        }
    }
```

2) 案例效果

应用式(5.18)对如图 5-9(a)所示的一个顶点在原点的长方体关于 x 轴进行对称变换，最终效果如图 5-9(b)所示。

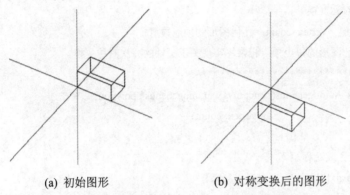

(a) 初始图形　　　　　　(b) 对称变换后的图形

图 5-9　关于 x 轴的齐次对称变换示意图

关于 x 轴的齐次对称

2. 关于 y 轴的齐次对称变换

三维图形关于 y 轴做对称变换的参数化表达式如下：

$$\begin{cases} x' = -x \\ y' = y \\ z' = -z \end{cases} \tag{5.19}$$

对应的齐次对称变换的矩阵乘积形式如下：

$$\begin{pmatrix} x' \\ y' \\ z' \\ 1 \end{pmatrix} = \begin{pmatrix} -1 & 0 & 0 & 0 \\ 0 & 1 & 0 & 0 \\ 0 & 0 & -1 & 0 \\ 0 & 0 & 0 & 1 \end{pmatrix} \begin{pmatrix} x \\ y \\ z \\ 1 \end{pmatrix} \tag{5.20}$$

1) 变换函数代码实现

代码如下：

```
/*************************************
*关于 y 轴的齐次对称变换
```

* polygon：几何图形；vertex_count：几何图形的顶点数量
 * 将对称矩阵与几何图形的每个顶点相乘，实现关于 y 轴的齐次对称变换
 **************************************/
void my_reflection_y_homogeneous(struct my_v_homogeneous* polygon,
                                 int vertex_count)
{
    //将公式(5.20)装配生成对称矩阵
    float translate_matrix[4][4];
    memset(translate_matrix, 0, sizeof(int) * 16);
    translate_matrix[0][0] = -1;
    translate_matrix[1][1] = 1;
    translate_matrix[2][2] = -1;
    translate_matrix[3][3] = 1;

    //遍历三维图形的每个顶点并对称多边形的每个顶点
    for (int vIndex = 0; vIndex < vertex_count; vIndex++)
    {
        struct my_v_homogeneous input_v;
        input_v.x = polygon[vIndex].x;
        input_v.y = polygon[vIndex].y;
        input_v.z = polygon[vIndex].z;
        input_v.ratio = 1;
        input_v = matrix_multiply_vector(translate_matrix, input_v);
        polygon[vIndex].x = input_v.x;
        polygon[vIndex].y = input_v.y;
        polygon[vIndex].z = input_v.z;
        polygon[vIndex].ratio = input_v.ratio;
    }
}
```

2) 案例效果

应用式(5.20)对如图 5-10(a)所示的一个顶点在原点的长方体关于 $y$ 轴进行齐次对称变换，最终效果如图 5-10(b)所示。
```

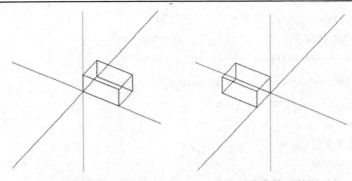

关于 y 轴的齐次对称

(a) 初始图形　　　　(b) 对称变换后的图形

图 5-10　关于 y 轴的齐次对称变换示意图

3. 关于 z 轴的齐次对称变换

三维图形关于 z 轴做对称变换的参数化表达式如下：

$$\begin{cases} x' = -x \\ y' = -y \\ z' = z \end{cases} \tag{5.21}$$

对应的齐次对称变换的矩阵乘积形式如下：

$$\begin{pmatrix} x' \\ y' \\ z' \\ 1 \end{pmatrix} = \begin{pmatrix} -1 & 0 & 0 & 0 \\ 0 & -1 & 0 & 0 \\ 0 & 0 & 1 & 0 \\ 0 & 0 & 0 & 1 \end{pmatrix} \begin{pmatrix} x \\ y \\ z \\ 1 \end{pmatrix} \tag{5.22}$$

1) 变换函数代码实现

代码如下：

```
/**********************************
* 关于 z 轴的齐次对称变换
* polygon：几何图形；vertex_count：几何图形的顶点数量
* 将对称矩阵与几何图形的每个顶点相乘，实现关于 z 轴的齐次对称变换
**********************************/
void my_reflection_z_homogeneous(struct my_v_homogeneous* polygon,
                                 int vertex_count)
{
    //将式(5.22)装配生成对称矩阵
    float translate_matrix[4][4];
    memset(translate_matrix, 0, sizeof(int) * 16);
```

```
translate_matrix[0][0] = -1;
translate_matrix[1][1] = -1;
translate_matrix[2][2] = 1;
translate_matrix[3][3] = 1;

//遍历三维图形的每个顶点并对称多边形的每个顶点
for (int vIndex = 0; vIndex < vertex_count; vIndex++)
{
    struct my_v_homogeneous input_v;
    input_v.x = polygon[vIndex].x;
    input_v.y = polygon[vIndex].y;
    input_v.z = polygon[vIndex].z;
    input_v.ratio = 1;
    input_v = matrix_multiply_vector(translate_matrix, input_v);
    polygon[vIndex].x = input_v.x;
    polygon[vIndex].y = input_v.y;
    polygon[vIndex].z = input_v.z;
    polygon[vIndex].ratio = input_v.ratio;
}
}
```

2) 案例效果

应用式(5.22)对如图 5-11(a)所示的一个顶点在原点的长方体关于 z 轴进行齐次对称变换，最终效果如图 5-11(b)所示。

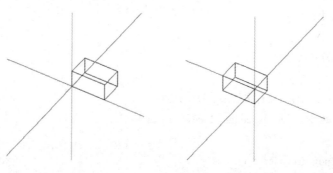

关于 z 轴的
齐次对称

(a) 初始图形　　　　(b) 对称变换后的图形

图 5-11　关于 z 轴的齐次对称变换示意图

4. 关于 *xoz* 坐标轴平面的齐次对称变换

三维图形关于 *xoz* 坐标轴平面做对称变换的参数化表达式如下：

$$\begin{cases} x' = x \\ y' = -y \\ z' = z \end{cases} \tag{5.23}$$

对应的齐次对称变换的矩阵乘积形式如下：

$$\begin{pmatrix} x' \\ y' \\ z' \\ 1 \end{pmatrix} = \begin{pmatrix} 1 & 0 & 0 & 0 \\ 0 & -1 & 0 & 0 \\ 0 & 0 & 1 & 0 \\ 0 & 0 & 0 & 1 \end{pmatrix} \begin{pmatrix} x \\ y \\ z \\ 1 \end{pmatrix} \tag{5.24}$$

1) 变换函数代码实现

代码如下：

```
/**************************************
 * 关于 xoz 坐标轴平面的齐次对称变换
 * polygon：几何图形；vertex_count：几何图形的顶点数量
 * 将对称矩阵与几何图形的每个顶点相乘，实现关于 xoz 坐标轴平面的齐次对称变换
 **************************************/
void my_reflection_xoz_homogeneous(struct my_v_homogeneous* polygon, int vertex_count)
{
    //将式(5.24)装配生成对称矩阵
    float translate_matrix[4][4];
    memset(translate_matrix, 0, sizeof(int) * 16);
    translate_matrix[0][0] = 1;
    translate_matrix[1][1] = -1;
    translate_matrix[2][2] = 1;
    translate_matrix[3][3] = 1;

    //遍历三维图形的每个顶点并对称多边形的每个顶点
    for (int vIndex = 0; vIndex < vertex_count; vIndex++)
    {
        struct my_v_homogeneous input_v;
        input_v.x = polygon[vIndex].x;
        input_v.y = polygon[vIndex].y;
```

```
            input_v.z = polygon[vIndex].z;
            input_v.ratio = 1;
            input_v = matrix_multiply_vector(translate_matrix, input_v);
            polygon[vIndex].x = input_v.x;
            polygon[vIndex].y = input_v.y;
            polygon[vIndex].z = input_v.z;
            polygon[vIndex].ratio = input_v.ratio;
        }
    }
```

2) 案例效果

应用式(5.24)对如图 5-12(a)所示的一个顶点在原点的长方体关于 xoz 坐标轴平面进行齐次对称变换，最终效果如图 5-12(b)所示。

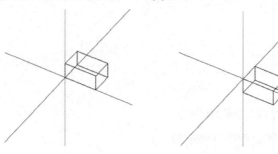

(a) 初始图形　　　　　(b) 对称变换后的图形

图 5-12　关于 xoz 坐标轴平面的齐次对称变换示意图

关于 xoz 坐标轴平面的齐次对称

5. 关于 xoy 坐标轴平面的齐次对称变换

三维图形关于 xoy 坐标轴平面做对称变换的参数化表达式如下：

$$\begin{cases} x' = x \\ y' = y \\ z' = -z \end{cases} \tag{5.25}$$

对应的齐次对称变换的矩阵乘积形式如下：

$$\begin{pmatrix} x' \\ y' \\ z' \\ 1 \end{pmatrix} = \begin{pmatrix} 1 & 0 & 0 & 0 \\ 0 & 1 & 0 & 0 \\ 0 & 0 & -1 & 0 \\ 0 & 0 & 0 & 1 \end{pmatrix} \begin{pmatrix} x \\ y \\ z \\ 1 \end{pmatrix} \tag{5.26}$$

1) 变换函数代码实现

代码如下:

```
/*****************************************
* 关于 xoy 坐标轴平面的齐次对称变换
* polygon: 几何图形; vertex_count: 几何图形的顶点数量
* 将对称矩阵与几何图形的每个顶点相乘,实现关于 xoy 坐标轴平面的齐次对称变换
*****************************************/
void my_reflection_xoy_homogeneous(struct my_v_homogeneous* polygon,
                                   int vertex_count)
{
    //将式(5.26)装配生成对称矩阵
    float translate_matrix[4][4];
    memset(translate_matrix, 0, sizeof(int) * 16);
    translate_matrix[0][0] = 1;
    translate_matrix[1][1] = 1;
    translate_matrix[2][2] = -1;
    translate_matrix[3][3] = 1;

    //遍历三维图形的每个顶点并对称多边形的每个顶点
    for (int vIndex = 0; vIndex < vertex_count; vIndex++)
    {
        struct my_v_homogeneous input_v;
        input_v.x = polygon[vIndex].x;
        input_v.y = polygon[vIndex].y;
        input_v.z = polygon[vIndex].z;
        input_v.ratio = 1;
        input_v = matrix_multiply_vector(translate_matrix, input_v);
        polygon[vIndex].x = input_v.x;
        polygon[vIndex].y = input_v.y;
        polygon[vIndex].z = input_v.z;
        polygon[vIndex].ratio = input_v.ratio;
    }
}
```

2) 案例效果

应用式(5.26)对如图 5-13(a)所示的一个顶点在原点的长方体关于 xoy 坐标轴平面进行齐次对称变换，最终效果如图 5-13(b)所示。

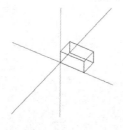
(a) 初始图形

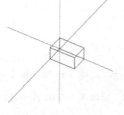
(b) 对称变换后的图形

图 5-13　关于 xoy 坐标轴平面的齐次对称变换示意图

关于 xoy 坐标轴平面的齐次对称

6. 关于 yoz 坐标轴平面齐次对称变换

三维图形关于 yoz 坐标轴平面做对称变换的参数化表达式如下：

$$\begin{cases} x' = -x \\ y' = y \\ z' = z \end{cases} \tag{5.27}$$

对应的齐次对称变换的矩阵乘积形式如下：

$$\begin{pmatrix} x' \\ y' \\ z' \\ 1 \end{pmatrix} = \begin{pmatrix} -1 & 0 & 0 & 0 \\ 0 & 1 & 0 & 0 \\ 0 & 0 & 1 & 0 \\ 0 & 0 & 0 & 1 \end{pmatrix} \begin{pmatrix} x \\ y \\ z \\ 1 \end{pmatrix} \tag{5.28}$$

1) 变换函数代码实现

代码如下：

```
/*********************************
 * 关于 yoz 坐标轴平面的齐次对称变换
 * polygon：几何图形；vertex_count：几何图形的顶点数量
 * 将对称矩阵与几何图形的每个顶点相乘，实现关于 yoz 坐标轴平面的齐次对称变换
 *********************************/
void my_reflection_yoz_homogeneous(struct my_v_homogeneous* polygon,
                                   int vertex_count)
{
    //将式(5.28)装配生成对称矩阵
```

```
float translate_matrix[4][4];
memset(translate_matrix, 0, sizeof(int) * 16);
translate_matrix[0][0] = -1;
translate_matrix[1][1] = 1;
translate_matrix[2][2] = 1;
translate_matrix[3][3] = 1;

//遍历三维图形的每个顶点并对称多边形的每个顶点
for (int vIndex = 0; vIndex < vertex_count; vIndex++)
{
    struct my_v_homogeneous input_v;
    input_v.x = polygon[vIndex].x;
    input_v.y = polygon[vIndex].y;
    input_v.z = polygon[vIndex].z;
    input_v.ratio = 1;
    input_v = matrix_multiply_vector(translate_matrix, input_v);
    polygon[vIndex].x = input_v.x;
    polygon[vIndex].y = input_v.y;
    polygon[vIndex].z = input_v.z;
    polygon[vIndex].ratio = input_v.ratio;
}
}
```

2) 案例效果

应用式(5.28)对如图 5-14(a)所示的一个顶点在原点的长方体关于 yoz 坐标轴平面进行齐次对称变换，最终效果如图 5-14(b)所示。

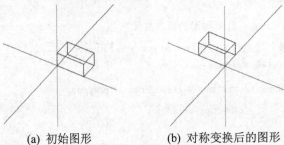

(a) 初始图形　　　　(b) 对称变换后的图形

图 5-14　关于 yoz 坐标轴平面齐次对称变换示意图

关于 yoz 坐标轴平面的齐次对称

5.3 三维复合几何变换

5.3.1 三维图形绕空间任意轴旋转

三维图形绕任意轴旋转的效果等价于多个三维基本几何变换的有序复合。此处，假定要实现任一三维图形绕过点 $P(x, y, z)$ 且方向向量为 $n(n_x, n_y, n_z)$ 的轴逆时针旋转 θ，其复合变换公式如下：

$$\begin{pmatrix} x' \\ y' \\ z' \\ 1 \end{pmatrix} = GFEDCBA \begin{pmatrix} x \\ y \\ z \\ 1 \end{pmatrix} \tag{5.29}$$

其中，A 为平移矩阵，对应式(5.2)，使三维图形及旋转轴在 x 轴、y 轴和 z 轴方向上平移 $(-x, -y, -z)$，达到 P 点与原点重合；B 为旋转矩阵，对应式(5.6)，使三维图形及旋转轴绕 y 轴逆时针旋转 α，达到旋转轴与 xoy 坐标轴平面重合；C 为旋转矩阵，对应式(5.8)，使三维图形及旋转轴绕 z 轴逆时针旋转 β，达到旋转轴与 x 轴重合；D 为旋转矩阵，对应式(5.4)，使得三维图形绕 x 轴逆时针旋转 θ；E 为旋转矩阵 C 的逆变换，对应式(5.8)，使三维图形及旋转轴绕 z 轴顺时针旋转 β，达到旋转轴与 xoy 坐标轴平面再次重合；F 为旋转矩阵 B 的逆变换，对应式(5.6)，使三维图形及旋转轴绕 y 轴顺时针旋转 α，达到旋转轴回到初始的姿态；G 为平移矩阵 A 的逆变换，对应式(5.2)，使得三维图形及旋转轴在 x 轴、y 轴和 z 轴方向上平移 (x, y, z)，达到 P 点与回到原来的位置。

1. 变换函数代码实现

代码如下：

```
/***********************************
* 绕任意轴旋转的函数
* polygon：几何图形；vertex_count：几何图形的顶点数量；x、y、z：任意轴上一个点的坐标；
* u、v、w：任意轴的方向向量；angle：几何图形的旋转角度
************************************/
void my_rotateArbitraryAxis_homogeneous(struct my_v_homogeneous* polygon,
            int vertex_count, float x, float y, float z, float u, float v, float w, float angle)
{
    //计算平移到原点的平移量
```

```
float tx = -x;
float ty = -y;
float tz = -z;
float pi = 3.1415926;

float angle1 = atan2(w, u) * 180 / pi;
float angle2 = atan2(v, sqrt(u * u + w * w)) * 180 / pi;    //向量与面的夹角的求法

//平移至与原点重合
my_translate_homogeneous(polygon, vertex_count, tx, ty, tz);
//将旋转轴逆时针绕 y 轴旋转到 xoy 坐标轴平面
my_rotate_y_homogeneous(polygon, vertex_count, angle1);
//将旋转轴顺时针绕 z 轴旋转到与 x 轴重合
my_rotate_z_homogeneous(polygon, vertex_count, -angle2);
//将旋转轴逆时针绕 x 轴旋转 angle 角度
my_rotate_x_homogeneous(polygon, vertex_count, angle);

//执行逆运算
my_rotate_z_homogeneous(polygon, vertex_count, angle2);
my_rotate_y_homogeneous(polygon, vertex_count, -angle1);
my_translate_homogeneous(polygon, vertex_count, -tx, -ty, -tz);
}
```

2. 案例效果

采用绕任意轴旋转算法，对如图 5-15(a)所示的一顶点在原点的长方体，关于绕过点 (20，20，20)且方向向量为(1，1，1)的轴逆时针旋转 60°，最终效果如图 5-15(b)所示。

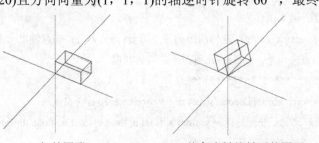

(a) 初始图形　　　　(b) 绕任意轴旋转后的图形

图 5-15　三维图形绕空间任意轴旋转示意图

三维图形绕空间
任意轴旋转

5.3.2 三维图形相对任意点缩放

三维图形相对任意点缩放的效果等价于多个三维基本几何变换的有序复合。此处，假定要实现任一三维图形相对于点 $P(x, y, z)$ 进行缩放，其复合变换公式如下：

$$\begin{pmatrix} x' \\ y' \\ z' \\ 1 \end{pmatrix} = CBA \begin{pmatrix} x \\ y \\ z \\ 1 \end{pmatrix} \tag{5.30}$$

其中：A 为平移矩阵，对应式(5.2)，使三维图形及 P 在 x 轴、y 轴和 z 轴方向上平移 $(-x, -y, -z)$，达到 P 点与原点重合；B 为缩放矩阵，对应式(5.10)，使三维图形绕原点缩放；C 为平移矩阵 A 的逆变换，对应式(5.2)，使得三维图形及 P 在 x 轴、y 轴和 z 轴方向上平移(x, y, z)，达到 P 点与回到原来的位置。

1. 变换函数代码实现

代码如下：

```
/*********************************
 * 相对任意点缩放的函数
 * polygon：几何图形；vertex_count：几何图形的顶点数量；x、y、z：三维空间任意一点；
 * sx、sy、sz：x 轴、y 轴、z 轴方向上的缩放系数
 **********************************/
void my_scaleArbitraryPoint_homogeneous(struct my_v_homogeneous* polygon,
                    int vertex_count, float x, float y, float z, float sx, float sy, float sz)
{
    //计算平移量
    float tx = -x;
    float ty = -y;
    float tz = -z;

    my_translate_homogeneous(polygon, vertex_count, tx, ty, tz);        //进行平移变换
    my_scale_homogeneous(polygon, vertex_count, sx, sy, sz);            //进行比例缩放
    my_translate_homogeneous(polygon, vertex_count, -tx, -ty, -tz);     //平移回原来的位置
}
```

2. 案例效果

采用绕任意点缩放算法，对如图 5-16(a)所示的一顶点在原点的长方体关于点(−100，−120，−220)进行缩放，设置其在 x、y、z 轴上的缩放系数分别为 0.6、0.8、1.2，最终效果如图 5-16(b)所示。

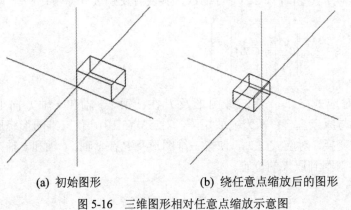

三维图形相对
任意点缩放

(a) 初始图形　　　　　　(b) 绕任意点缩放后的图形

图 5-16　三维图形相对任意点缩放示意图

5.4　三维几何变换综合示例

本书使用三个长方体，模拟手臂的多层次变换，实现绕公共轴旋转。设计步骤如下：
(1) 绘制出有共同边的三个长方体。
(2) 实现基础的三维物体变换。
(3) 用基础变换实现三维物体绕任意轴旋转(平移运动、绕任意轴旋转的运动)。
(4) 根据需要将绕任意轴旋转的变换通过键盘交互事件施加在相应物体上。
(5) 观察实验结果，看是否达到模拟手臂的效果。

1. 关键步骤

(1) 计算公共轴的方向向量，代码如下：

```
float x = boxA[3].x - boxA[0].x;
float y = boxA[3].y - boxA[0].y;
float z = boxA[3].z - boxA[0].z;
```

(2) 调用 5.3.1 小节中的绕任意轴旋转函数，实现长方体 B 绕公共轴旋转，代码如下：

```
my_rotateArbitaryAxis_homogeneous(boxB, 8, boxA[0].x, boxA[0].y, boxA[0].z, boxA[3].x,
                                  boxA[3].y, boxA[3].z, 45);
```

2. 案例效果

应用上述三维几何变换算法,实现模拟手臂变换的三个长方体的最终效果如图 5-17 所示。

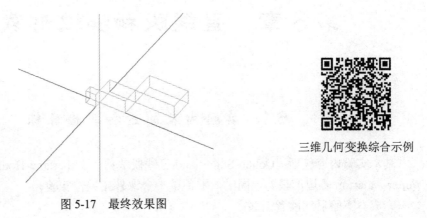

三维几何变换综合示例

图 5-17 最终效果图

5.5 课外拓展性实验

任务 1:通过向量 t(4,-3,8),将一点 p_1(-2,6,3)进行平移,求目标点 p_2 在齐次坐标系下的坐标。

任务 2:有一个四面体 $A_1B_1C_1D_1$,其中 A_1、B_1、C_1、D_1 的顶点坐标分别为(0,0,0)、(1,0,0)、(0,1,0)和(0,0,1)。将该四面体绕过点 C_1 且方向向量为(0,1,1)的直线逆时针方向旋转 45°,求旋转之后的顶点 A_1、B_1、D_1。

第6章 直线段和多边形裁剪

6.1 实验内容简述和实验目标

基本实验内容包括：Cohen-Sutherland 直线段裁剪、Sutherland-Hodgman 多边形裁剪、Weiler-Atherton 多边形裁剪。同时，配备了一个课外拓展性实验。

完成本实验后，读者能够：

(1) 熟记和描述 Cohen-Sutherland 直线段裁剪的原理和过程(布鲁姆知识模型：记忆和理解)；

(2) 熟记和描述 Sutherland-Hodgman 多边形裁剪的原理和过程(布鲁姆知识模型：记忆和理解)；

(3) 熟记和描述 Weiler-Atherton 多边形裁剪的原理和过程(布鲁姆知识模型：记忆和理解)；

(4) 判断交点是出点还是入点(布鲁姆知识模型：应用)；

(5) 判断点(累积角度法)是在多边形内部还是在多边形外部(布鲁姆知识模型：应用)；

(6) 结合 OpenGL 编程实现——Cohen-Sutherland 直线段裁剪、Sutherland-Hodgman 多边形裁剪和 Weiler-Atherton 多边形裁剪(布鲁姆知识模型：应用)。

6.2 Cohen-Sutherland 直线段裁剪

Cohen-Sutherland 直线段裁剪算法的核心思想是：将每条线段 P_1P_2 分为以下三种情况来处理。

(1) 若线段 P_1P_2 完全在裁剪多边形内，则完整保留该线段。

(2) 若线段 P_1P_2 明显在裁剪多边形外，则直接丢弃该线段。

(3) 若线段 P_1P_2 不满足(1)、(2)条件，则求出线段与裁剪多边形的交点，将该线段分成两段，丢弃完全在裁剪多边形外的线段，然后对另一段迭代上述处理。

为了快速判断一条直线段与裁剪多边形的位置关系，可采用的编码方法是：延长裁剪多边形的边，将二维平面分成 9 个区域，给每个区域赋予 4 位编码(见图 6-1)。

裁剪一条线段时(见图 6-2)，先求出线段两端点 P_1、P_4 的编码 code1、code2。若 code1=0、code2=0，则线段 P_1P_2 完全在裁剪多边形内，直接完整保留该线段。若编码按位与运算 code1&code2≠0，则说明两个端点同在裁剪多边形的上方、下方、左方或右方，可判断线段完全在裁剪多边形外，丢弃整条线段；否则求出线段与裁剪多边形的交点，将直线段分成两段，先丢弃完全在裁剪多边形外的线段，再对剩下的另一段线段迭代开展上述操作。

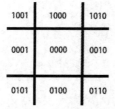

图 6-1　多边形裁剪区域编码

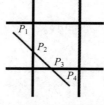

图 6-2　线段裁剪

1. 关键数据结构

自定义如下数据结构用以表示三维齐次坐标下的点(顶点或者交点)。该数据结构应用于本书随后所有章节的实验中。

```
struct my_homogeneous_point
{
    float x;
    float y;
    float z=0;
    float ratio;
};
```

2. 关键函数代码实现

代码如下：

```
/*********************************
* 编码函数，输入点坐标，输出 4 位编码
* px、py：输入点；clip_polygon：裁剪多边形；
* clip_polygon_point_count：裁剪多边形的顶点数量
* 4 位编码的意义如下(从右到左)：
* 第 1 位：如果端点在裁剪多边形的左侧，则为 1，否则为 0
* 第 2 位：如果端点在裁剪多边形的右侧，则为 1，否则为 0
```

* 第 3 位：如果端点在裁剪多边形的下侧，则为 1，否则为 0
* 第 4 位：如果端点在裁剪多边形的上侧，则为 1，否则为 0
***/

```
my_homogeneous_point clip_polygon[clip_polygon_point_count];
unsigned int enCode(double px, double py)
{
    //用 1 2 4 8 编码裁剪多边形的左、右、下、上区域
    unsigned int RC = 0;
    if (px < clip_polygon[0].x)
    {
        RC = RC | 1;
    }
    else if (px > clip_polygon[1].x)
    {
        RC = RC | 2;
    }
    if (py < clip_polygon[0].y)
    {
        RC = RC | 4;
    }
    else if (py > clip_polygon[3].y)
    {
        RC = RC | 8;
    }
    return RC;
}

//Cohen-Sutherland 裁剪函数
//根据线段的两个端点 point1 和 point2 的编码判断该线段与裁剪多边形的位置，求出
//裁剪后的两个端点
void cilpping_lines(my_homogeneous_point point1，my_homogeneous_point point2)
{
    float x，y;                              //储存变换后的 x，y 坐标
```

```
int code1 = enCode(point1.x, point1.y);        //code1、code2 记录顶点的编码
int code2 = enCode(point2.x, point2.y);
int code = 0;

while (code1 != 0 || code2 != 0)               //两直线至少有一点在区域外
{
    if ((code1 & code2) != 0) return;          //两点连线不过裁剪区域,跳出循环

    //将在区域外的点编码给 code
    code = code1;
    if (code1 == 0) code = code2;

    if ((1 & code) != 0)                       //直线段与左边界相交,求出直线与边界的相交点
    {
        x = clip_polygon[0].x;
        y = point1.y + (point2.y - point1.y) *
            (clip_polygon[0].x - point1.x) / (point2.x - point1.x);
    }
    else if ((2 & code) != 0)                  //直线段与右边界相交,求出直线与边界的相交点
    {
        x = clip_polygon[1].x;
        y = point1.y + (point2.y - point1.y) *
            (clip_polygon[1].x - point1.x) / (point2.x - point1.x);
    }
    else if ((4 & code) != 0)                  //直线段与下边界相交,求出直线与边界的相交点
    {
        y = clip_polygon[0].y;
        x = point1.x + (point2.x - point1.x) *
            (clip_polygon[0].y - point1.y) / (point2.y - point1.y);
    }
    else if ((8 & code) != 0)                  //直线段与上边界相交,求出直线与边界的相交点
    {
        y = clip_polygon[3].y;
```

```
            x = point1.x + (point2.x - point1.x) *
            (clip_polygon[3].y - point1.y) / (point2.y - point1.y);
        }

        //将求出的点重新赋值
        if (code == code1)
        {
            point1.x = x;
            point1.y = y;
            code1 = enCode(x, y);
        }
        else
        {
            point2.x = x;
            point2.y = y;
            code2 = enCode(x, y);
        }
    }

    //将新生成的点存入 List 中
    List.push_back(point1);
    List.push_back(point2);
}
```

3. 案例效果

图 6-3(a)显示了直线和裁剪多边形的初始状态，平移裁剪多边形，用 Cohen-Sutherland 算法执行裁剪，最终效果如图 6-3(b)所示。

(a) 初始状态　　　　(b) 裁剪多边形移动后的状态

Cohen-Sutherland
直线段裁剪

图 6-3　用 Cohen-Sutherland 算法执行裁剪示意图

6.3 Sutherland-Hodgman 多边形裁剪

Sutherland-Hodgman 多边形裁剪算法的基本思想是：遍历裁剪多边形的每一条边，用逐条边去裁剪被裁剪多边形。因此，在算法的每一步中，仅需考虑裁剪多边形的一条边及其延长线构成的裁剪线。该裁剪线把平面分成两个部分：可见一侧(即包含裁剪多边形的一侧)和不可见一侧(即不包含裁剪多边形的一侧)。按序考虑被裁剪多边形每条边端点 S、P 与裁剪线的位置关系，有以下四种情况：

(1) 若 S、P 均在可见一侧，则保留点 P(见图 6-4(a))。
(2) 若 S、P 均在不可见一侧，则保留 0 个点(见图 6-4(b))。
(3) 若 S 可见，P 不可见，则保留 SP 与裁剪线的交点 I(见图 6-4(c))。
(4) 若 S 不可见，P 可见，则保留 SP 与裁剪线的交点 I 和点 P(见图 6-4(d))。

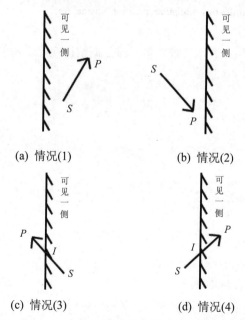

图 6-4　S、P 与裁剪线的四种位置关系

上述算法仅用一条裁剪边对多边形进行裁剪，可得到一个顶点序列，将该序列作为下一条裁剪边处理的输入。

1. 关键数据结构

自定义点相对于多边形的空间位置关系，其数据结构如下：

```
enum relative_position
{
    inside_on,      //内部
    outside         //外部
};
```

自定义如下数据结构用以表示扩展的三维齐次坐标下的点(顶点或者交点)。该数据结构应用于本书随后所有章节的实验中。

```
struct my_homogeneous_point_EX
{
    float x;
    float y;
    float z=0;
    float ratio;
    enum relative_position pos = relative_position::inside_on;
};
```

自定义如下数据结构用以表示三维向量。该数据结构应用于本书随后所有章节的实验中。

```
class my_3Dvector
{
    public:
    float dx;
    float dy;
    float dz;
    float len;
    public:
    my_3Dvector();
    my_3Dvector(float x, float y, float z);
    //start 点指向 end 点的向量
    my_3Dvector(my_homogeneous_point_EX start, my_homogeneous_point_EX end);
    my_3Dvector cross_multiply(my_3Dvector input_vector);      //向量叉乘
    float dot_multiply(my_3Dvector input_vector);              //向量点乘
};
```

2. 关键函数代码实现

代码如下：

```
/*****************************************
 * 两条边求交点函数
 * a1、a2：一条边上的两个端点；b1、b2：另一条边上的两个端点
 *****************************************/
struct my_homogeneous_point_EX GetCrossPoint(
                    my_homogeneous_point_EX a1, my_homogeneous_point_EX a2,
                    my_homogeneous_point_EX b1, my_homogeneous_point_EX b2)
{
    float A1 = a2.y - a1.y;
    float B1 = a1.x - a2.x;
    float C1 = a2.x * a1.y - a1.x * a2.y;
    float A2 = b2.y - b1.y;
    float B2 = b1.x - b2.x;
    float C2 = b2.x * b1.y - b1.x * b2.y;
    my_homogeneous_point_EX cross;
    cross.x = (C2 * B1 - C1 * B2) / (A1*B2-A2*B1);
    cross.y = (A1*C2-A2*C1)/(A2*B1-A1*B2);
    return cross;
}


/*****************************************
 * 判断给定点是在给定裁剪边的可见一侧还是不可见一侧
 * edge_start_point、edge_end_point：当前裁剪边的起点和终点；given_point：给定点；
 * clip_polygon：裁剪多边形
 *****************************************/
my_homogeneous_point_EX clip_polygon[clip_polygon_point_count];
bool determine_point_edge_position(my_homogeneous_point_EX edge_start_point,
                    my_homogeneous_point_EX edge_end_point,
                    my_homogeneous_point_EX given_point)
{
    //求出裁剪平面上相邻的两个向量，计算裁剪平面(即投影面)的法向量
    my_3Dvector a(clip_polygon[1], clip_polygon[2]);
    my_3Dvector b(clip_polygon[1], clip_polygon[3]);
```

```
        my_3Dvector face_normal= a.cross_multiply(b);

    //通过叉乘计算相邻两个向量所形成的给定点的位置法向量
        my_3Dvector c(edge_start_point, edge_end_point);    //c 为裁剪线段向量
        my_3Dvector d(edge_start_point,given_point);        //d 为判断点与裁剪线段
        my_3Dvector cross_vector = c.cross_multiply(d);
    //通过点乘，判断给定点的位置法向量与裁剪平面的法向量是否同向
    //点乘值大于 0，视为同向，给定点在可见侧；否则为逆向，给定点在不可见侧
        float sign = cross_vector.dot_multiply(face_normal);
        return sign >= 1e-6;                                //在可见一侧返回 1,在不可见一侧返回 0;
    }


    /*******************************
    * 裁剪函数 Sutherland_Hodgman_Clipping
    * clip_polygon、subject_polygon：裁剪多边形和被裁剪多边形；
    * clip_polygon_point_count：裁剪多边形顶点个数；
    * subject_polygon_point_count：被裁剪多边形顶点个数
    *********************************/
    my_homogeneous_point_EX clip_polygon[clip_polygon_point_count];
    my_homogeneous_point_EX subject_polygon[subject_polygon_point_count];
    void Sutherland_Hodgman_Clipping()
    {
        for (int i = 0; i < clip_polygon_point_count; i++)
        {
            //第一步：取出当前点 clip_cur_point 和下一点 clip_next-point
            ListA.clear();      // ListA 为链表，用于存储每条边的裁剪结果
            int start_vec_index = i % clip_polygon_point_count;
            int end_vec_index = (i + 1) % clip_polygon_point_count;
            clip_cur_point = clip_polygon[start_vec_index];
            clip_next_point = clip_polygon[end_vec_index];
            bool flag = true;
            //第二步：遍历 subject_polygon，取出当前点 subject_cur_point 和下一点
            //subject_next_point
```

```
for (iter = ListB.begin(); flag == true; iter++)
{
    subject_cur_point = *iter;
    iter++;
    if (iter == ListB.end())
    {
        flag = false;
        iter = ListB.begin();
        subject_next_point = *iter;
    }
    else
    {
        subject_next_point = *iter;
    }
    if (iter != ListB.begin())
    {
        iter--;
    }

    bool a =determine_point_edge_position(clip_cur_point, clip_next_point, subject_cur_point);
    bool b=determine_point_edge_position(clip_cur_point,clip_next_point, subject_next_point);

    //根据规则，进行边求交或顶点取舍，如有点需要保留，则插入到 ListA 中
    if (a == false && b == true)
    {
        clip_cross_subject = GetCrossPoint(clip_cur_point, clip_next_point,
                             subject_cur_Point, subject_next_Point);
        ListA.push_back(clip_cross_subject);
        ListA.push_back(subject_next_Point);
    }
    else if (a == true && b == true)
    {//都在可见一侧，储存终点
        ListA.push_back(subject_next_Point);
```

```
            }
            else if (a == true && b == false)
            {//起点在可见一侧，终点在不可见一侧，储存交点
                clip_cross_subject = GetCrossPoint(clip_cur_point, clip_next_point,
                                                    subject_cur_Point, subject_next_Point);
                ListA.push_back(clip_cross_subject);
            }
        }

        //第三步：将 ListA 复制到 ListB 数组，再回到第一步
        //ListB 用于存储最终的裁剪结果
        for (iter = ListA.begin(); iter != ListA.end(); iter++)
        {
            ListB.push_back(*iter);
        }
    }
}
```

3. 案例效果

图 6-5(a)给定了裁剪多边形(粗线)和被裁剪多边形(细线)的初始状态，用 Sutherland-Hodgman 算法执行裁剪，最终效果如图 6-5(b)所示。

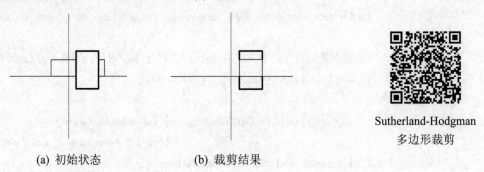

(a) 初始状态　　　　　　(b) 裁剪结果

图 6-5 用 Sutherland-Hodgman 算法执行裁剪示意图

Sutherland-Hodgman
多边形裁剪

6.4　Weiler-Atherton 多边形裁剪

Weiler-Atherton 多边形裁剪算法能有效开展凹多边形的裁剪。假设被裁剪多边形和裁

剪多边形的顶点都按逆时针方向排列。当上述两个多边形相交时，交点通常成对出现，即存在多对"入点"和"出点"，沿着被裁剪多边形的边逆时针游走，通过入点进入裁剪多边形内部；沿着被裁剪多边形的边逆时针游走，通过出点离开裁剪多边形内部。根据上述现象，Weiler-Atherton 多边形裁剪算法从被裁剪多边形中选择一个未被访问的入点，沿着该点所在的边逆时针游走。在此过程中，若遇到入点，则算法继续沿着被裁剪多边形的边逆时针游走；若遇到出点，则沿着裁剪多边形的边逆时针游走。游走过程直至遇到刚开始的入点时停止，即找到了一个裁剪结果区域。迭代上述过程，直至所有入点都被访问，即找到了所有裁剪结果区域。其算法步骤如下：

(1) 定义裁剪多边形 clip_polygon 和被裁剪多边形 subject_polygon，它们均为 my_homogeneous_point_EX 数组。同时，定义三个带头指针的循环链表 plist、clip_polygonlist 和 subject_polygonlist，分别用来存储交点、插入交点的裁剪多边形和插入交点的被裁剪多边形。

(2) 定义一个多边形求交函数 generateIntersectPoints，实现两个多边形的所有交点存储在 plist 链表中，在此过程中，将上述每一交点按序(相交边的两端点之间)插入 subject_polygonlist 和 clip_polygonlist。

(3) 定义函数用来确定每个交点为出点或入点。

(4) 定义生成所有裁剪结果区域的函数 generateClipArea，其具体步骤如下：

① 定义记录裁剪结果的临时数组 subject_polygonClip。

② 从头指针开始顺序遍历 subject_polygonlist，寻找第一个未被访问过的入点，若存在，则将该入点设为当前点，并记录到 subject_polygonClip 中，进入步骤③，否则进入步骤⑤。

③ 从当前点开始遍历 subject_polygonlist，将下一个被访问的点 PC 设为当前点，并记录到 subject_polygonClip 中。若 PC 为 subject_polygonClip 的第一个点，则将 subject_polygonClip 数据输出，即找到一个裁剪结果区域，之后清空 subject_polygonClip，并进入步骤②；而若 PC 不是出点，则继续步骤③操作，否则在 clip_polygonlist 中找到 PC，并进入步骤④。

④ 从当前点开始遍历 clip_polygonlist，将下一个被访问的点 PN 设为当前点，并记录到 subject_polygonClip 中。若 PN 不是入点，则继续步骤④操作，否则进入步骤③。

⑤ 退出函数，裁剪完成。

1. 关键数据结构

自定义以下数据结构，用于存储两个多边形相交的交点：

```
struct IntersectionPoint
{
    my_homogeneous_point_EX p;
    int pointFlag;        //标记是否为交点：1 为交点，0 为原有点
```

```
    int index0, index1;    //点在裁剪多边形第 index0 条线段上，在被裁剪多边形第 index1 条线段上
    bool inFlag;           //出点为 false，其他点默认为 true
    int distance;          //交点距离所在线段起点的距离，用于在交点排序的时候使用
};
```

2. 关键代码实现

代码如下：

```cpp
void generateClipAreas(list<IntersectionPoint>& clip_polygonlist,
                       list<IntersectionPoint>& subject_polygonlist,
                       list<IntersectionPoint>& plist)
{
    list<IntersectionPoint>::iterator it1, it2;
    list<IntersectionPoint>::iterator it3;                    //it3 用来比较交点是否都被遍历
    vector<my_homogeneous_point_EX> subject_polygonClip;      //临时裁剪结果
    vector<my_homogeneous_point_EX> allpoints;                //用来记录所有遍历过的点
    bool visitall = false;

    //在被裁剪多边形 subject_polygonlist 中找到交点且为入点的第一个点
    for (it2 = subject_polygonlist.begin(); it2 != subject_polygonlist.end(); it2++)
        if (it2->pointFlag == 1 && it2->inFlag) break;

    subject_polygonClip.pts.clear();
    while (visitall==false)           //若 plist 中所有交点被访问，则跳出循环
    {
        visitall = true;
        for (it3 = plist.begin(); it3 != plist.end(); it3++)
        {
            bool flag = false;        //遍历 plist 中的每一个点，观察是否被裁剪
            if (allpoints.pts.size() == 0)
            {
                visitall = false;
            }
            for (int i = 0; i < allpoints.pts.size(); i++)
            {
```

```
        //如果找到对应点
        if ((it3->p.x == allpoints.pts[i].x) && (it3->p.y == allpoints.pts[i].y))
        {
            flag = true;
        }

        if (flag == false)
        {
            if (i == allpoints.pts.size() - 1)        //遍历到最后还没找到对应点
            {
                visitall = false;
            }
        }
    }
}

//全都被访问,直接跳出循环
if (visitall == true)
{
    continue;
}

//循环遍历 subject_polygonlist
it2 = (it2 == subject_polygonlist.end()) ? subject_polygonlist.begin() : it2;

subject_polygonClip.pts.push_back(it2->p);
allpoints.pts.push_back(it2->p);
it2++;
it2 = (it2 == subject_polygonlist.end()) ? subject_polygonlist.begin() : it2;

for (; it2 != subject_polygonlist.end(); it2++)
{
    //subject_polygonlist 遍历到结尾,要从头开始
```

```
            if (it2->pointFlag == 1 && !it2->inFlag) break;
            subject_polygonClip.pts.push_back(it2->p);
            allpoints.pts.push_back(it2->p);
        }

        //在 clip_polygonlist 中找同一点
        for (it1 = clip_polygonlist.begin(); it1 != clip_polygonlist.end(); it1++)
            if (it1->p == it2->p) break;

        for (; it1 != clip_polygonlist.end(); it1++)
        {
            if (it1->pointFlag == 1 && it1->inFlag) break;
            subject_polygonClip.pts.push_back(it1->p);
            allpoints.pts.push_back(it1->p);
        }

        //首尾点相等，找到一个裁剪区域，输出，并查找下一个裁剪区域
        if (circle(subject_polygonClip, it1))
        {
            subject_polygonClip.drawPg();      //输出当前找到的裁剪区域
            continue;                          //回到最外层 while，判断 while (visitall==false)
        }

        //在 subject_polygonlist 中找同一点
        for (; it2 != subject_polygonlist.end(); it2++)
            if (it2->p == it1->p) break;
    }
}
```

3. 案例效果

如图 6-6(a)所示为裁剪多边形(粗线)和被裁剪多边形(细线)的初始状态，用 Weiler-Atherton 算法执行裁剪，裁剪最终效果如图 6-6(b)所示。

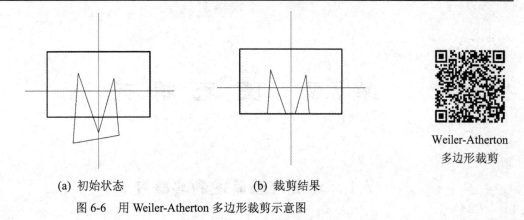

(a) 初始状态　　　　　　　　(b) 裁剪结果

图 6-6　用 Weiler-Atherton 多边形裁剪示意图

Weiler-Atherton
多边形裁剪

6.5　课外拓展性实验

如图 6-7 所示为裁剪多边形(粗线)和被裁剪多边形(细线)的初始状态，对其执行 Weiler-Atherton 多边形裁剪。

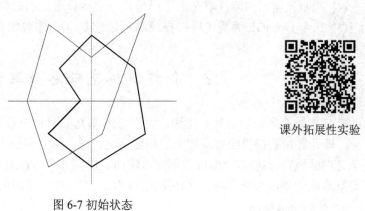

图 6-7　初始状态

课外拓展性实验

第 7 章 图元填充

7.1 实验内容简述和实验目标

基本实验内容为扫描线填充任意多边形。同时，配备了一个课外拓展性实验。
完成本实验后，读者能够：
(1) 熟记和描述光栅多边形填充原理(布鲁姆知识模型：记忆和理解)；
(2) 熟记和描述扫描线填充多边形的原理和过程(布鲁姆知识模型：记忆、理解)；
(3) 熟记和描述种子点填充多边形区域的原理和过程(布鲁姆知识模型：记忆、理解)；
(4) 判断光栅扫描转换填充多边形过程中边界异常点的取舍(布鲁姆知识模型：应用)；
(5) 结合 OpenGL 编程实现扫描线填充任意多边形(布鲁姆知识模型：应用)。

7.2 扫描线填充任意多边形

扫描线填充多边形的基本思想：将多边形填充分解到一系列连续的线扫描求交和填充上来，即计算每条扫描线与多边形的相交情况，进而确定多边形内部区域的直线段(内部需要填充的区域)，实现多边形内部填充。然而，直接开展上述求交填充的运算量非常大，为了提高算法效率，需充分利用多边形的边界(边)连贯性，采用增量的方式进行填充。

1. 关键数据结构

为了准确记录和分类每条边与扫描线的相交情况，使用了如下多边形边(Edge)的数据结构：

```
struct Edge
{
    float cur_X;            //当前扫描线与当前边交点的 x 坐标值
    float reciprocal_K;     //当前边斜率的倒数
    int max_Y;              //当前边两端点最大的 y 坐标值
```

```
Edge(int y, float x, float rec_k)
{
    max_Y = y;
    cur_X = x;
    reciprocal_K = rec_k;
}
```
};

  同时，为了充分利用多边形边的连贯性或相邻关系，采用常见的边表(ET)来分类保存多边形的所有边——按照每条边的两端点 y 坐标最小值对多边形的非水平边进行分类表示的指针数组，有多少条扫描线，ET 就有多少分类，同一类中的所有边先按照 x 值递增排序，其次按边的 reciprocal_K 递增排序。

  随着扫描线顺序更迭，为了能快速确定哪些边与当前扫描线有交点(避免真正开展求交计算)，采用了活化边表(AET)，用来把与当前扫描线存在相交情况的所有边进行统一处理，同时依据它们与扫描线的交点坐标的 x 值，按递增的顺序存放。

2. 算法关键步骤

(1) 生成多边形对应的 ET 表。

(2) 将扫描线纵坐标 y 的初值置为 ET 中非空元素的最小序号。

(3) 置 AET 为空。

(4) 执行下列步骤直至 ET 和 AET 都为空：

① 如 ET 中的第 y 类非空，则将其中的所有边取出并插入 AET 中。

② 新边插入 AET 时与 AET 中原有边重新排序。

③ 对 AET 中的边两两配对，将每对边中的 x 坐标按规则取整，获得有效的填充区段，将 AET 每条边中的 x 递增 reciprocal_K。

④ 将当前扫描线纵坐标 y 值递增 1。

⑤ 将 AET 中满足 y 项与递增后的 y 值相同的边删去。

⑥ 迭代步骤④，直到 y 达到最大值。

3. 关键代码实现

代码如下：

```
/*************************************
 * 生成多边形的 ET 表
 * polygon：逆时针排列的顶点数组
 * 遍历 polygon，相邻两顶点之间构造一条边 Edge，并同时根据两顶点 y 坐标的最小值，对所有边
 * 进行分类和插入排序
```

**************************************/
```cpp
map<int, vector<Edge>> ET;                    //根据 y 值对所有边进行分类存放
void generate_ET()
{
    for (int eindex = 0; eindex < polygon.size(); eindex++)
    {
        my_v_inhomogeneous startPoint = polygon[eindex];
        my_v_inhomogeneous endPoint = (eindex == (polygon.size() - 1)) ? polygon[0] :
                                                        polygon[eindex + 1];
        if (endPoint.y == startPoint.y) continue;      //水平边不加入 ET 表

        //生成一条边
        my_v_inhomogeneous top_point, bottom_point;
        top_point = startPoint.y > endPoint.y ? startPoint : endPoint;
        bottom_point = startPoint.y < endPoint.y ? startPoint : endPoint;
        float rec_k = (top_point.x - bottom_point.x) / (top_point.y - bottom_point.y);
        Edge one_ET_Edge(top_point.y, bottom_point.x, rec_k);

        //以边两端点 y 坐标的最小值为依据对边进行分类和插入排序
        map<int, vector<Edge>>::iterator it = ET.find(bottom_point.y);
        if (it == ET.end())
        {
            vector<Edge> new_edge_list;
            new_edge_list.push_back(one_ET_Edge);
            ET.insert(pair<int, vector<Edge>>(bottom_point.y, new_edge_list));
        }
        else
        {
            bool inserted = false;
            vector<Edge>::iterator eit = it->second.begin();
            for (;eit != it->second.end(); eit++)
            {
                if (bottom_point.x < eit->cur_X ||
```

```
                    (bottom_point.x == eit->cur_X && rec_k < eit->reciprocal_K))
                {
                    it->second.insert(eit, one_ET_Edge);
                    inserted = true;
                    break;
                }
            }
            if (false == inserted)
            {
                it->second.push_back(one_ET_Edge);
            }
        }
    }
}

/***********************************
 * 往 AET 中插入新的 Edge 集合，并排序
 * polygon：逆时针排列的顶点数组
 ***********************************/
void insert_new_edges_into_AET(vector<Edge> newedges)
{
    //根据 AET 中边的排序规则，对每条边进行插入排序
    vector<Edge>::iterator newHead = newedges.begin();
    for (; newHead != newedges.end(); newHead++)
    {
        bool inserted = false;
        vector<Edge>::iterator AET_Head = AET.begin();
        for (;AET_Head != AET.end(); AET_Head++)
        {
            if (newHead->cur_X < AET_Head->cur_X ||
                (newHead->cur_X == AET_Head->cur_X &&
                newHead->reciprocal_K < AET_Head->reciprocal_K))
            {
```

```
                AET.insert(AET_Head, *newHead);
                inserted = true;
                break;
            }
        }
        if (false == inserted)
        {
            AET.push_back(*newHead);
        }
    }
}


/*****************************************
 * 扫描线填充
 *****************************************/
vector<Edge> AET;                                    //活化边表
void scanline_filling()
{
    AET.clear();
    for (int ypos = y_min; ypos <= y_max; ypos += 1)  //逐行遍历所有扫描线
    {
        //找到所有两端点 y 坐标的最小值为与当前扫描线 ypos 相同的所有边
        map<int, vector<Edge>>::iterator it = ET.find(ypos);
        if (it != ET.end())
        {
            insert_new_edges_into_AET(it->second);    //插入到 AET 表中
        }

        vector<int> need_removed_edge_index;          //存放需要删除的边序号

        //确定填充线段区域并填充
        glBegin(GL_LINES);
            for (int eindex = 0; eindex < AET.size(); eindex += 2)
```

```
                {
                    if (AET[eindex].reciprocal_K > 1)
                    {
                        glVertex2f(AET[eindex].cur_X + 2, ypos);
                    }
                    else
                    {
                        glVertex2f(AET[eindex].cur_X + 1, ypos);
                    }
                    AET[eindex].cur_X += 1 * AET[eindex].reciprocal_K;
                    glVertex2f(AET[eindex + 1].cur_X - 1, ypos);

                    AET[eindex + 1].cur_X += 1 * AET[eindex + 1].reciprocal_K;

                    if (AET[eindex].max_Y == ypos + 1)
                        need_removed_edge_index.push_back(eindex);

                    if (AET[eindex + 1].max_Y == ypos + 1)
                        need_removed_edge_index.push_back(eindex + 1);
                }
            glEnd();

            //删除 AET 中所有 max_Y == ypos 的所有边
            for (int eindex = need_removed_edge_index.size() - 1; eindex >= 0; eindex--)
            {
                AET.erase(AET.begin() + need_removed_edge_index[eindex]);
            }
        }
    }
```

4. 案例效果

如图 7-1(a)所示为未进行填充的多边形，执行扫描线填充后，最终效果如图 7-1(b)所示。

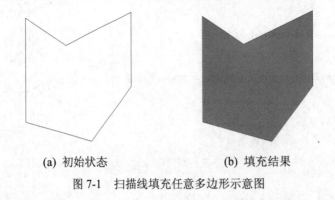

(a) 初始状态　　　　　(b) 填充结果

图 7-1　扫描线填充任意多边形示意图

扫描线填充任意多边形

7.3　课外拓展性实验

运用扫描线填充任意多边形算法对如图 7-2 所示的图形进行填充。

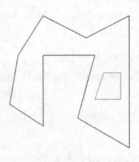

课外拓展性实验

图 7-2　未填充多边形

第 8 章 几何图元消隐

8.1 实验内容简述和实验目标

基本实验内容包括：z 缓冲消隐、背面剔除消隐。同时，配备了一个课外拓展性实验。完成本实验后，读者能够：
(1) 熟记和描述图像空间消隐和景物空间消隐的原理(布鲁姆知识模型：记忆、理解)；
(2) 熟记和描述 z 缓冲消隐的原理和过程(布鲁姆知识模型：记忆、理解)；
(3) 熟记和描述背面剔除消隐的原理和计算过程(布鲁姆知识模型：记忆、理解)；
(4) 结合 OpenGL 能实现背面剔除消隐和 z 缓冲消隐(布鲁姆知识模型：应用)。

8.2 z 缓冲消隐

z 缓冲算法也叫深度缓冲器算法，通常包括帧缓冲器和 z 缓冲两部分(见图 8-1)。在 OpenGL 里分别对应以下两个矩阵。
(1) Intensity(x, y)：属性矩阵(帧缓冲器)，存储每个像素的光强或颜色。
(2) Depth(x, y)：深度矩阵(z 缓冲)，存放每个像素对应图形上的点相对于视点的距离值，默认为该点的 z 值。

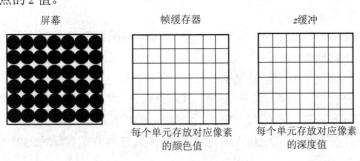

图 8-1 z 缓冲算法存储图示

如图 8-2 所示，假定 xoy 坐标轴平面为投影面，z 轴负方向为观察方向。过投影面上任意一点(x, y)，对应屏幕上的一个像素点，作平行于 z 轴的射线 R，与图形表面分别相交于 p_1 和 p_2。p_1 点和 p_2 点的 z 值称为该点的深度值。z 缓冲算法比较 p_1 点和 p_2 点的 z 值，将最大的 z 值存入 z 缓冲中。显然，p_1 点相对于 p_2 点更加接近视点，z 值更大，因此，(x, y)这一点将取图形 p_1 点的颜色。

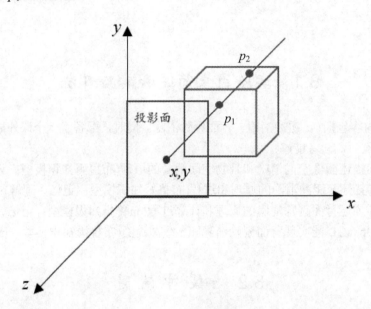

图 8-2　深度判断图示

1. 实验过程描述

本书取一个平行于 xoy 坐标轴平面作为投影面，视点放置于世界坐标系 z 轴正向的某个位置上。视点坐标系则是架设在视点上的标准右手正交坐标系，它由三个相互垂直的单位向量 u、v、n 组成。其中：向量 n 垂直于投影面，代表视线方向；向量 v 平行于世界坐标系中的 y 轴；向量 $u = n \times v$。

本书先将 z 缓冲中各单元的深度值初始化为 $-\infty$。当准备在某个像素点绘制某个投影点的颜色时，首先将该投影点的深度值和该像素点当前的深度值进行比较。若投影点的深度值更大，则用它的颜色替换像素原来的颜色，否则丢弃。

在 OpenGL 平台上，通过 glEnable 和 glDepthFunc 两个函数实现场景根据深度信息进行动态消隐。glEnable(GL_DEPTH_TEST)允许 OpenGL 开启深度测试判图形之间沿视线方向的遮挡关系或深度关系；glDepthFunc 设置目标像素与当前像素颜色的替代规则，其输入参数及含义如表 8-1 所示。

表 8-1　glDepthFunc 输入参数及含义

可取的参数值	含　　义
GL_NEVER	待绘制的深度值不取代储存在 z 缓冲中的深度值
GL_LESS	如果待绘制的深度值小于储存在 z 缓冲中的深度值，则通过
GL_LEQUAL	如果待绘制的深度值小于或等于储存在 z 缓冲中的深度值，则通过
GL_EQUAL	如果待绘制的深度值等于储存在 z 缓冲中的深度值，则通过
GL_GREATER	如果待绘制的深度值大于储存在 z 缓冲中的深度值，则通过
GL_NOTEQUAL	如果待绘制的深度值不等于储存在 z 缓冲中的深度值，则通过
GL_GEQUAL	如果待绘制的深度值大于或等于储存在 z 缓冲中的深度值，则通过
GL_ALWAYS	待绘制的深度值取代储存在 z 缓冲中的深度值

2. 案例效果

如图 8-3(a)所示，球的中心在原点，半径为 30；长方体的一个顶点与原点重合，长宽高分别为 80、40、50。在这里使用 glDepthFunc(GL_LEQUAL)，最终效果如图 8-3(b)所示。

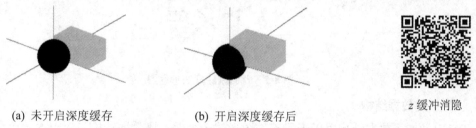

(a) 未开启深度缓存　　　　(b) 开启深度缓存后

z 缓冲消隐

图 8-3　开启深度缓存效果图

图 8-3(a)中未开启深度缓存，球和长方体之间的深度关系比较晦涩，沿视线方向未展现出交叉干涉效果或正确的遮挡效果。开启深度缓存后，图 8-3(b)中球与长方体交叉干涉和遮挡清晰明确。在未开启深度缓存时，窗口展现的内容会随着三维图形绘制先后顺序的不同而不同，即对于存在空间重叠的两个三维图形，先绘制的三维图形往往会被后绘制的三维图形所遮挡，如图 8-4 所示。通过深度缓存，能避免上述情况，无论绘制顺序如何，效果始终如一，如图 8-3(b)所示。

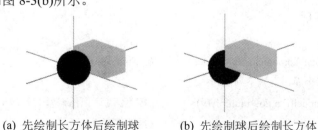

(a) 先绘制长方体后绘制球　　　　(b) 先绘制球后绘制长方体

图 8-4　三维图形绘制先后顺序不同效果图

8.3 背面剔除消隐

三维图形通常由很多离散的(网格)面构成。为了提高图形的绘制效率，在背面剔除算法中，背对视点的面(即背面)会被剔除出投影操作，仅确保面向视点的面。为了判断一个面是否需要剔除，通常依据视线方向 $v(v_x, v_y, v_z)$ 与该面的法向量 $n(n_x, n_y, n_z)$ 所成角度进行确定。如图 8-5 所示，令 θ 为视线方向 v 和法向量 n 的夹角。当夹角 $\theta \in [-90°, 90°]$ 时，即两向量的余弦值大于 0，则表明面背对视点，需要被剔除。否则，当 $\theta \in [-180°, -90°)$ 或 $\theta \in (90°, 180°]$ 时，即两向量的余弦值小于等于 0，则表明是可见面，应该被绘制。

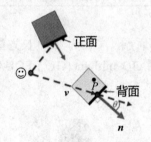

图 8-5 背面剔除算法示意图

1. 关键数据结构

自定义如下数据结构用以表示三维图形的一个面。该数据结构应用于本章后续所有相关实验。

```
struct my_face_homogeneous
{
    list<my_homogeneous_point> mList;    //各个顶点按照逆时针的顺序储存
    my_3Dvector n;                        //定义面法向量
};
```

2. 关键代码实现

代码如下：

```
vector<my_face_homogeneous> model;        //三维图形
//遍历模型每个面，实施下列操作：
float num = model[i].n.dot_multiply(v);   //模型第 i 个面法向量与视向量点乘
if (num > 0)                              //绘制可见面
{
```

```
glBegin(GL_POLYGON);
    list<my_homogeneous_point>::iterator iter = model[i].mList.begin();
    for (; iter != model[i].mList.end(); iter++)
    {
        glVertex3f((*iter).x, (*iter).y, (*iter).z);
    }
glEnd();
}
```

3. 案例效果

采用背面剔除算法绘制图 8-6(a)对应的三维图形(封闭)，并给每个面填上颜色便于观察，最终效果如图 8-6(b)所示。

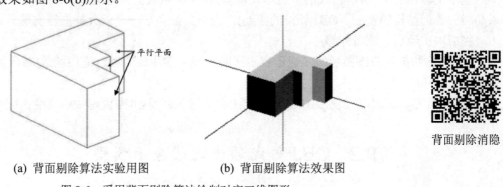

(a) 背面剔除算法实验用图　　　　(b) 背面剔除算法效果图

图 8-6　采用背面剔除算法绘制对应三维图形

背面剔除在一些情况下并不能正确剔除不可见面。如图 8-6(a)所示，箭头所指的三个右侧面法向量相等，根据背面剔除算法，这三个平面会同时显示，而绘制效果则由系统默认采用的深度排序算法决定。

8.4　课外拓展性实验

在将如图 8-6(a)所示的三维模型每个表面填上不同的颜色后，变换视点位置，观察图形上的面消隐是否全部正确，若不是，请找出那些消隐错误的区域。

第 9 章 三维图形的表示和加载

9.1 实验内容简述和实验目标

基本实验内容为 OBJ 格式的多边形表示模型(OBJ 格式文件含义、OBJ 模型内容的读入与相关数据结构的装配、OBJ 模型的绘制)。同时，配备了一个课外拓展性实验。

完成本实验后，读者能够：

(1) 熟记和描述三维图形的多边形表示法的原理、常用格式以及适用场景(布鲁姆知识模型：记忆和理解)；

(2) 结合 OpenGL 能实现 OBJ 格式三维图形的读入和显示(布鲁姆知识模型：应用)。

9.2 OBJ 格式的多边形表示模型

9.2.1 OBJ 格式文件含义

在三维图形的多边形表示中，本书载入以 OBJ 格式保存的模型。OBJ 文件的具体内容如图 9-1～图 9-4 所示，依次为顶点坐标表、顶点法向表、纹理坐标表和面表。通过 C++ 的 fstream 库中的函数，可以读取上述内容，装配成三维图形，并在 OpenGL 中绘制。

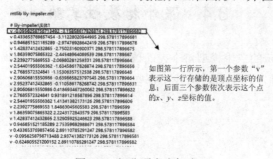

图 9-1 部分顶点坐标表

第 9 章 三维图形的表示和加载

```
vn 0.139015730053865 -0.919038702070925 -0.034168895333.2928
vn -0.452616521552778 -0.891597169334715 -0.013884237529955B
vn -0.457070741418588 -0.889346996155254 -0.012176114679502B
vn -0.509127830508061 -0.860639583983912 0.009399929901552B
vn -0.536621602243587 -0.843533570783683 0.022099116870470T
vn -0.563841925836935 -0.825112817893717 0.035652775712209B
vn -0.612227949583732 -0.788213500072057 0.062421278845102A
vn -0.616254226641956 -0.784875366073699 0.064818113800095T
vn -0.665914061585973 -0.73971459756359 0.096854409984311A
vn -0.682937046119994 -0.722295426018967 0.109024348606062
vn -0.712229933801253 -0.689484453736918 0.131680330545484
vn -0.745461003348756 -0.646690390394282 0.160647992187865
vn -0.754085114311434 -0.63469684166911 0.16886550727223
vn -0.797440643897402 -0.56335283318844 0.216152734486615
vn -0.821659359709504 -0.51276617696575 0.248891029101099
vn -0.837935599425712 -0.47127470674515 0.275252760199306
vn -0.864085105442932 -0.37672311093775 0.33382125192154
vn -0.864972572311914 -0.37221692977841 0.33656720627759
vn -0.876814938665852 -0.27115967289505 0.39707429216898
vn -0.877441419698051 -0.23448227839576 0.41852359219554
vn -0.873701639962029 -0.16862005876685 0.45630332029046
vn -0.861526027847967 -0.09166570682785 0.49936990451193
vn -0.855632958191729 -0.06714376531857 0.51287587431233
vn -0.824655538678933 0.02913244102175 0.56484539889589
vn -0.819058571283876 0.04276439250676 0.57211385539938
```

图 9-2 部分顶点法向表

如图第一行所示，第一个参数"vn"表示这一行存储的是顶点法向的信息；后面三个参数依次表示这个顶点法向的 x、y、z 的值。

```
vt 0.0120424243687753 0.0170826583183608
vt 0.0120424243687753 0.0190485536230037
vt 0.00292901283428878 0.0190485536230037
vt -0.00618439870011977 0.0190485536230037
vt -0.00618439870011977 0.0210144489276466
vt -0.00618439870011977 0.0200315012753252
vt 0.00292901283428878 0.0200315012753252
vt -0.00618439870011977 0.0170826583183608
vt -0.00618439870011977 0.0180656059706822
vt 0.00292901283428878 0.0180656059706822
vt 0.0120424243687753 0.0131508677090175
vt 0.0120424243687753 0.0151167630137179
vt 0.00292901283428878 0.0151167630137179
vt -0.00618439870011977 0.0131508677090175
vt -0.00618439870011977 0.0151167630137179
vt 0.00292901283428878 0.00921907709978914
vt 0.0120424243687753 0.00921907709978914
vt 0.0120424243687753 0.0111849724044321
vt 0.00292901283428878 0.0111849724044321
vt -0.00618439870011977 0.0111849724044321
vt -0.00162769293295446 0.0111849724044321
vt -0.00162769293295446 0.0131508677090175
vt -0.00162769293295446 0.0121679200567535
vt -0.00618439870011977 0.0121679200567535
vt 0.00292901283428878 0.0121679200567535
```

图 9-3 部分纹理坐标表

如图第一行所示，第一个参数"vt"表示这一行存储的是纹理坐标的信息；后面两个参数依次表示这个点对应的纹理坐标的 u、v 值。

```
f 3893/3893/3893 3900/3900/3900 3892/3892/3892
f 3892/3892/3892 3900/3900/3900 3899/3899/3899
f 3892/3892/3892 3899/3899/3899 3896/3896/3896
f 3896/3896/3896 3899/3899/3899 3902/3902/3902
f 3845/3845/3845 3844/3844/3844 3880/3880/3880
f 3880/3880/3880 3844/3844/3844 3879/3879/3879
f 3880/3880/3880 3879/3879/3879 3881/3881/3881
f 3881/3881/3881 3879/3879/3879 3878/3878/3878
f 3881/3881/3881 3878/3878/3878 3883/3883/3883
f 3883/3883/3883 3878/3878/3878 3882/3882/3882
f 3883/3883/3883 3882/3882/3882 3884/3884/3884
f 3884/3884/3884 3882/3882/3882 3877/3877/3877
f 3884/3884/3884 3877/3877/3877 3894/3894/3894
f 3894/3894/3894 3877/3877/3877 3893/3893/3893
f 3894/3894/3894 3893/3893/3893 3895/3895/3895
f 3895/3895/3895 3893/3893/3893 3892/3892/3892
f 3895/3895/3895 3892/3892/3892 3897/3897/3897
f 3897/3897/3897 3892/3892/3892 3896/3896/3896
f 3847/3847/3847 3888/3888/3888 3907/3907/3907
f 3907/3907/3907 3888/3888/3888 3906/3906/3906
f 3907/3907/3907 3906/3906/3906 3905/3905/3905
f 3888/3888/3888 3876/3876/3876 3906/3906/3906
f 3906/3906/3906 3876/3876/3876 3885/3885/3885
f 3847/3847/3847 3907/3907/3907 3898/3898/3898
f 3898/3898/3898 3907/3907/3907 3891/3891/3891
```

如图第一行所示，第一个参数"f"表示这一行存储的是面的信息；后面三个参数依次表示这个三角面片三个顶点的信息，每个参数以"v/vt/vn"的索引组成。
如图所示，第一行的含义为：这个三角面片由第3893、第3900、第3892这三个顶点组成，平面的纹理由第3893、第3900、第3892这三个纹理坐标形成，这个平面的朝向是第3893、第3900、第3892这三个顶点的法向量求平均值确定。

图 9-4 部分面表

9.2.2 OBJ 模型内容的读入与相关数据结构的装配

OBJ 模型内容读入的伪代码如下：

遍历 OBJ 文件的每一行，并进行如下具体分析：

{

  1. 若该行以"v"字段开头，则表示该行为顶点信息，读取该行中剩余的三个浮点数字段，分别对应该顶点的 x、y 和 z 坐标，并装配成一个顶点数据结构，存储到 OBJ 模型对应的顶点坐标序列中。

  2. 若该行以"vn"字段开头，则表示该行为单位法向信息，读取该行中剩余的三个浮点数字段，分别对应该法向在 x、y 和 z 坐标下的分量，并装配成一个法向数据结构，存储到 OBJ 模型对应的法向序列中。

  3. 若该行以"vt"字段开头，则表示该行为归一化的纹理坐标信息，读取该行中剩余的两个浮点数字段，分别对应该纹理坐标在 u 和 v 坐标下的分量，并装配成一个纹理坐标数据结构，存储到 OBJ 模型对应的纹理坐标序列中。

  4. 若该行以"f"字段开头，则表示该行为面属性信息，读取该行中的剩余三个字符串，分别表示面上逆时针存放的三个顶点(序号)以及它们各自相关的纹理坐标(序号)和法向(序号)，将它们装配到面数据结构中，并存储到 OBJ 模型对应的面序列中。

}

1. 主要数据结构

(1) 顶点数据结构：

```
struct my_3D_point_coord
{
    float x;      //x 轴坐标值
    float y;      //y 轴坐标值
    float z;      //z 轴坐标值
};
```

(2) 纹理坐标数据结构：

```
struct my_2D_Texture_coord
{
    float u;      //u 方向上的坐标
    float v;      //v 方向上的坐标
};
```

(3) 面数据结构，即构成三维图形的一个三角网格面的关键数据结构：

```
struct my_triangle_indices
{
    int first_point_index;              //第一个点序号
    int first_point_texture_index;      //第一个纹理坐标序号
    int first_point_normal_index;       //第一个点法向序号

    int second_point_index;             //第二个点序号
    int second_point_texture_index;     //第二个纹理坐标序号
    int second_point_normal_index;      //第二个法向序号

    int third_point_index;              //第三个点序号
    int third_point_texture_index;      //第三个纹理坐标序号
    int third_point_normal_index;       //第三个法向序号
};
```

(4) 三维图形数据结构，即三角网格模型的关键数据结构：

```
struct my_triangle_3DModel
{
    …
    vector<my_3D_point_coord> pointSets;            //存放模型所有顶点
    vector<my_3Dvector> pointNormalSets;            //存放模型所有顶点的法向
    vector<my_2D_Texture_coord> pointTextureSets;   //存放模型所有纹理坐标
    vector<my_triangle_indices> faceSets;           //存放模型所有三角网格面
    …
};
```

2. 关键代码实现

代码如下：

```
my_triangle_3DModel my_3DModel;                     //三维网格模型
vector<string> parameters;
//读入OBJ文件的一行，根据空格进行分割，把每个字段存入parameters
if (parameters[0] == "v")                           //顶点
{
    my_3D_point_coord curPoint;
```

```cpp
        curPoint.x = atof(parameters[1].c_str());
        curPoint.y = atof(parameters[2].c_str());
        curPoint.z = atof(parameters[3].c_str());
        my_3DModel.pointSets.push_back(curPoint);
    }
    else if (parameters[0] == "vn")        //顶点的法向量
    {
        my_3Dvector curPointNormal;
        curPointNormal.dx = atof(parameters[1].c_str());
        curPointNormal.dy = atof(parameters[2].c_str());
        curPointNormal.dz = atof(parameters[3].c_str());
        my_3DModel.pointNormalSets.push_back(curPointNormal);
    }
    else if (parameters[0] == "vt")        //纹理坐标
    {
        my_2D_Texture_coord curTextureCoord;
        curTextureCoord.u = atof(parameters[1].c_str());
        curTextureCoord.v = atof(parameters[2].c_str());
        my_3DModel.pointTextureSets.push_back(curTextureCoord);
    }
    else if (parameters[0] == "f")         //面，顶点索引/纹理 uv 索引/法向索引
    {
        //顶点索引在 OBJ 文件中是从 1 开始，而顶点 vector 是从 0 开始，因此要减 1
        my_triangle_indices curTri;
        curTri.first_point_index = atoi(parameters[1].substr(0,
                            parameters[1].find_first_of('/')).c_str()) - 1;
        parameters[1] = parameters[1].substr(parameters[1].find_first_of('/') + 1);
        curTri.first_point_texture_index = atoi(parameters[1].substr(0,
                            parameters[1].find_first_of('/')).c_str()) - 1;
        parameters[1] = parameters[1].substr(parameters[1].find_first_of('/') + 1);
        curTri.first_point_normal_index = atoi(parameters[1].substr(0,
                            parameters[1].find_first_of('/')).c_str()) - 1;
```

第 9 章 三维图形的表示和加载

```
            curTri.second_point_index = atoi(parameters[2].substr(0,
                            parameters[2].find_first_of('/')).c_str()) - 1;
    parameters[2] = parameters[2].substr(parameters[2].find_first_of('/') + 1);
    curTri.second_point_texture_index = atoi(parameters[2].substr(0,
                            parameters[2].find_first_of('/')).c_str()) - 1;
    parameters[2] = parameters[2].substr(parameters[2].find_first_of('/') + 1);
    curTri.second_point_normal_index = atoi(parameters[2].substr(0,
                            parameters[2].find_first_of('/')).c_str()) - 1;

    curTri.third_point_index = atoi(parameters[3].substr(0,
                            parameters[3].find_first_of('/')).c_str()) - 1;
    parameters[3] = parameters[3].substr(parameters[3].find_first_of('/') + 1);
    curTri.third_point_texture_index = atoi(parameters[3].substr(0,
                            parameters[3].find_first_of('/')).c_str()) - 1;
    parameters[3] = parameters[3].substr(parameters[3].find_first_of('/') + 1);
    curTri.third_point_normal_index = atoi(parameters[3].substr(0,
                            parameters[3].find_first_of('/')).c_str()) - 1;
    my_3DModel.faceSets.push_back(curTri);
}
```

9.2.3 OBJ 模型的绘制

完成 OBJ 模型内容的读入与相关数据结构的装配，本书采用下面的算法进行模型绘制：

遍历装配完成的面序列(即 OBJ 模型)
{
  1. 根据当前面中的三个顶点序号，提取面的三个顶点坐标
  2. 根据当前面中的三个顶点纹理坐标序号，提取三个顶点纹理坐标
  3. 根据当前面中的三个顶点法向序号，提取三个顶点法向
  绘制
}

1. 关键代码实现

代码如下：

```
my_triangle_3DModel cur3DModel;          //当前装配完成的 OBJ 模型
for (int i = 0; i < cur3DModel.faceSets.size(); i++)
```

```
{
    //取出顶点序号获得相应顶点坐标
    my_3D_point_coord point1, point2, point3;
    int firstPointIndex = cur3DModel.faceSets[i].first_point_index;
    int secondPointIndex = cur3DModel.faceSets[i].second_point_index;
    int thirdPointIndex = cur3DModel.faceSets[i].third_point_index;
    point1 = cur3DModel.pointSets[firstPointIndex];        //第一个顶点
    point2 = cur3DModel.pointSets[secondPointIndex];       //第二个顶点
    point3 = cur3DModel.pointSets[thirdPointIndex];        //第三个顶点

    //取出法向序号获得相应法向
    my_3Dvector vector1, vector2, vector3;
    int firstNormalIndex = cur3DModel.faceSets[i].first_point_normal_index;
    int secondNormalIndex = cur3DModel.faceSets[i].second_point_normal_index;
    int thirdNormalIndex = cur3DModel.faceSets[i].third_point_normal_index;
    vector1 = cur3DModel.pointNormalSets[firstNormalIndex];    //第一个点的法向量
    vector2 = cur3DModel.pointNormalSets[secondNormalIndex];   //第二个点的法向量
    vector3 = cur3DModel.pointNormalSets[thirdNormalIndex];    //第三个点的法向量

    //取出纹理坐标序号获得相应纹理坐标
    my_2D_Texture_coord texture1, texture2, texture3;
    int firstTextureIndex = cur3DModel.faceSets[i].first_point_texture_index;
    int secondTextureIndex = cur3DModel.faceSets[i].second_point_texture_index;
    int thirdTextureIndex = cur3DModel.faceSets[i].third_point_texture_index;
    texture1 = cur3DModel.pointTextureSets[firstTextureIndex];    //第一个点的纹理
    texture2 = cur3DModel.pointTextureSets[secondTextureIndex];   //第二个点的纹理
    texture3 = cur3DModel.pointTextureSets[thirdTextureIndex];    //第三个点的纹理

    //绘制三角网格面
    glBegin(GL_TRIANGLES);
        glNormal3f(vector1.dx, vector1.dy, vector1.dz);        //绑定顶点的法向
        glTexCoord2f(texture1.u, texture1.v);                  //绑定顶点的纹理坐标
        glVertex3f(point1.x, point1.y, point1.z);              //绘制顶点
```

```
        glNormal3f(vector2.dx, vector2.dy, vector2.dz);
        glTexCoord2f(texture2.u, texture2.v);
        glVertex3f(point2.x, point2.y, point2.z);

        glNormal3f(vector3.dx, vector3.dy, vector3.dz);
        glTexCoord2f(texture3.u, texture3.v);
        glVertex3f(point3.x, point3.y, point3.z);
    glEnd();
}
```

2. 案例效果

将如图 9-5(a)所示的 OBJ 文件，根据上述代码导入到 OpenGL 环境中，在不考虑灯光、纹理、z 缓冲消隐时，效果如图 9-5(b)所示；加上 z 缓冲消隐和灯光后，效果如图 9-6 所示。

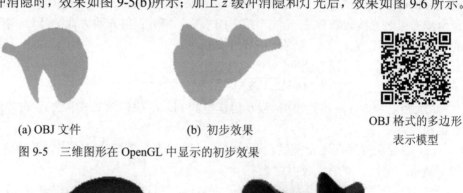

(a) OBJ 文件　　　　(b) 初步效果

图 9-5　三维图形在 OpenGL 中显示的初步效果

OBJ 格式的多边形表示模型

图 9-6　加上 z 缓冲消隐和灯光后三维图形在 OpenGL 中的显示效果

9.3　课外拓展性实验

编写代码，将更加复杂的 OBJ 格式文件(字段可以更加丰富)导入到 OpenGL 环境中，在不考虑灯光、纹理、z 缓冲消隐时，观察效果；对比加上 z 缓冲消隐和灯光后的效果。

第10章 光照计算

10.1 实验内容简述和实验目标

基本实验内容包括：直接光照计算(环境光、漫反射光、镜面反射光、Phong 光照模型、Blinn-Phong 光照模型)、明暗过渡计算(Flat 明暗过渡、Gouraud 明暗过渡、Phong 明暗过渡)及全局光照之光线跟踪算法(光线跟踪中的求交计算、反射光线方向的计算、折射光线方向的计算)。同时，配备了一个课外拓展性实验。

完成本实验后，读者能够：
(1) 熟记颜色模型(布鲁姆知识模型：记忆)；
(2) 熟记和描述环境光模型、漫反射模型和 Phong 模型的原理和公式(布鲁姆知识模型：记忆和理解)；
(3) 熟记和描述环境光模型、漫反射模型和 Phong 模型的优缺点以及适用范围(布鲁姆知识模型：记忆、理解、应用)；
(4) 熟记、描述和使用 Gouraud 以及 Phong 明暗过渡(布鲁姆知识模型：记忆、理解和应用)；
(5) 找到模型网格面明暗过渡和高光效果存在的问题，并提出改进策略(布鲁姆知识模型：应用、分析)；
(6) 结合 OpenGL 能实现环境光模型、漫反射模型、Phong 模型、Gouraud 明暗过渡以及 Phong 明暗过渡(布鲁姆知识模型：应用)。

10.2 直接光照计算

10.2.1 环境光

环境光是指光源所产生的光在环境中经过多次反射，最终达到平衡时分布在环境中的

一种光。可近似认为同一环境下的环境光,其光强分布是均匀的,计算公式如下:

$$I_{ambient} = I_a \times K_a \tag{10.1}$$

其中:I_a 表示环境光的能量值;K_a 表示物体表面对环境光的反射系数。

根据式(10.1),定义了两个数组 light_rgb_ambirnt 和 material_ambient_rgb_reflection,用于保存环境光的能量值和环境光的反射系数。每个数组各有三个元素,分别保存 R、G 和 B 三色的能量值。

1. 关键代码实现

代码如下:

```
float ambient_r = light_rgb_ambirnt[0] * material_ambient_rgb_reflection[0];
float ambient_g = light_rgb_ambirnt[1] * material_ambient_rgb_reflection[1];
float ambient_b = light_rgb_ambirnt[2] * material_ambient_rgb_reflection[2];
```

2. 案例效果

根据式(10.1)给加载的 OBJ 模型设置了环境光反射系数,并给场景设置了环境光,效果如图 10-1 所示。

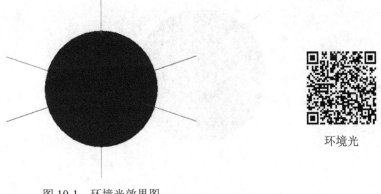

图 10-1 环境光效果图

10.2.2 漫反射光

当光线照射到比较粗糙的物体表面时,会形成漫反射。漫反射光的能量值近似服从 Lambert 定律,其计算公式如下:

$$I_{diffuse} = I_p \times K_d \times (L \cdot N) \tag{10.2}$$

其中:I_p 表示入射光的能量值;K_d 表示物体的表面漫反射率;L 表示光源向量;N 表示物

体光照点处的法向量；$L \cdot N$ 表示两向量的夹角(入射角)余弦值。

定义 light_rgb_diffuse_specular 数组保存入射光的能量值；定义 material_diffuse_rgb_reflection 数组保存物体表面的漫反射系数；定义 costheta 保存 $L \cdot N$ 的值。

1. 关键代码实现

代码如下：

```
float diffuse_r = costheta * light_rgb_diffuse_specular[0] *
        material_diffuse_rgb_reflection[0];
float diffuse_g = costheta * light_rgb_diffuse_specular[1] *
        material_diffuse_rgb_reflection[1];
float diffuse_b = costheta * light_rgb_diffuse_specular[2] *
        material_diffuse_rgb_reflection[2];
```

2. 案例效果

根据式(10.2)给 OBJ 模型设置漫反射系数，并给场景设置了点光源，最终效果如图 10-2 所示。

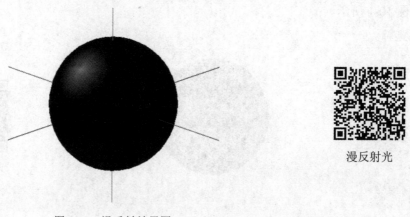

漫反射光

图 10-2　漫反射效果图

10.2.3　镜面反射光

光照射到相对光滑的物体表面往往会产生镜面反射，其特点是在光滑表面会产生高光区域，经验模型表达式如下：

$$I_{spec} = I_p \times K_s \times (R \cdot V)^n \tag{10.3}$$

其中：I_p 表示入射光的能量值；K_s 表示物体表面高光反射系数；R 表示反射光向量；V 表

示视线方向的逆向量(表示光线从物体表面光照点进入眼睛的方向)；n 表示高光指数，反映物体表面的光滑程度。

与 10.2.2 小节的漫反射光类似，这里依旧使用数组 light_rgb_diffuse_specular 来保存入射光的能量值；定义 material_specular_rgb_reflection 数组来保存物体表面高光反射系数；定义 cosalpha 来保存 $\boldsymbol{R} \cdot \boldsymbol{V}$ 的值。

1. 关键代码实现

代码如下：

```
float specular_r = light_rgb_diffuse_specular[0] * material_specular_rgb_reflection[0] *
        pow(cosalpha, n);
float specular_g = light_rgb_diffuse_specular[1] * material_specular_rgb_reflection[1] *
        pow(cosalpha, n);
float specular_b = light_rgb_diffuse_specular[2] * material_specular_rgb_reflection[2] *
        pow(cosalpha, n);
```

2. 案例效果

根据式(10.3)给 OBJ 模型设置镜面反射系数，并给场景设置了点光源，最终效果如图 10-3 所示。

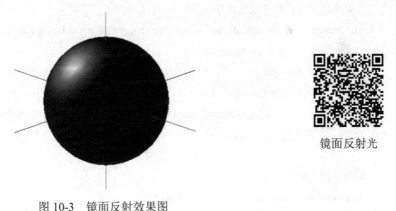

镜面反射光

图 10-3　镜面反射效果图

10.2.4　Phong 光照模型

在单一光源照射下，Phong 光照模型为环境光、漫反射光和镜面反射光之和，其表达式如下：

$$I = I_a \times K_a + I_p \times K_d \times (\boldsymbol{L} \cdot \boldsymbol{N}) + I_p \times K_s \times (\boldsymbol{R} \cdot \boldsymbol{V})^n \tag{10.4}$$

其中各参数的含义同前，此处不再赘述。

1. 关键代码实现

代码如下：

```
//环境光能量值计算
float ambient_r = light_rgb_ambirnt[0] * material_ambient_rgb_reflection[0];
float ambient_g = light_rgb_ambirnt[1] * material_ambient_rgb_reflection[1];
float ambient_b = light_rgb_ambirnt[2] * material_ambient_rgb_reflection[2];

//入射光的漫反射能量值计算
float diffuse_r = costheta * light_rgb_diffuse_specular[0] *
                  material_diffuse_rgb_reflection[0];
float diffuse_g = costheta * light_rgb_diffuse_specular[1] *
                  material_diffuse_rgb_reflection[1];
float diffuse_b = costheta * light_rgb_diffuse_specular[2] *
                  material_diffuse_rgb_reflection[2];

//入射光的镜面反射能量值计算
float specular_r = light_rgb_diffuse_specular[0] * material_specular_rgb_reflection[0] *
                   pow(cosalpha, n);
float specular_g = light_rgb_diffuse_specular[1] * material_specular_rgb_reflection[1] *
                   pow(cosalpha, n);
float specular_b = light_rgb_diffuse_specular[2] * material_specular_rgb_reflection[2] *
                   pow(cosalpha, n);

//得到总的能量值和应用
float total_r = ambient_r + diffuse_r + specular_r;
float total_g = ambient_g + diffuse_g + specular_g;
float total_b = ambient_b + diffuse_b + specular_b;
glColor3f(total_r, total_g, total_b);
```

2. 案例效果

给场景设置点光源，并使用式(10.4)给添加了环境光反射系数、漫射系数和镜面光反射系数的 OBJ 模型计算直接光照，最终效果如图 10-4 所示。

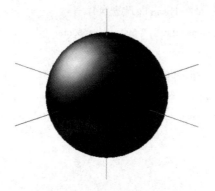

Phong 光照模型

图 10-4 Phong 光照模型效果图

10.2.5 Blinn-Phong 光照模型

根据式(10.4)可知，Phong 光照模型计算镜面光时，当 R 和 V 的夹角超过 90°时，它们的余弦值变为小于 0，这和自然现象不符合。针对该问题，本实验引入了如式(10.5)所示的 Blinn-Phong 光照模型。

$$I=I_a \times K_a + I_p \times K_d \times (L \Box N) + I_p \times K_s \times (N \Box H)^n \tag{10.5}$$

式(10.5)在 Phong 光照模型的基础上引入了向量 H，即 V 和 L 的中间向量，以避免高光部分出现负值。

1. 关键代码实现

代码如下：

```
//环境光能量值计算
float ambient_r = light_rgb_ambirnt[0] * material_ambient_rgb_reflection[0];
float ambient_g = light_rgb_ambirnt[1] * material_ambient_rgb_reflection[1];
float ambient_b = light_rgb_ambirnt[2] * material_ambient_rgb_reflection[2];

//入射光的漫反射能量值计算
float diffuse_r = costheta * light_rgb_diffuse_specular[0] *
                  material_diffuse_rgb_reflection[0];
float diffuse_g = costheta * light_rgb_diffuse_specular[1] *
                  material_diffuse_rgb_reflection[1];
float diffuse_b = costheta * light_rgb_diffuse_specular[2] *
                  material_diffuse_rgb_reflection[2];
```

//入射光的镜面反射能量值计算，此处 cosalpha 代表 N·H
float specular_r = light_rgb_diffuse_specular[0] * material_specular_rgb_reflection[0] *
    pow(cosalpha, n);
float specular_g = light_rgb_diffuse_specular[1] * material_specular_rgb_reflection[1] *
    pow(cosalpha, n);
float specular_b = light_rgb_diffuse_specular[2] * material_specular_rgb_reflection[2] *
    pow(cosalpha, n);

//得到总的能量值和应用
float total_r = ambient_r + diffuse_r + specular_r;
float total_g = ambient_g + diffuse_g + specular_g;
float total_b = ambient_b + diffuse_b + specular_b;
glColor3f(total_r, total_g, total_b);

2. 案例效果

给场景设置点光源，并使用式(10.5)给添加了环境光反射系数、漫射系数和镜面光反射系数的 OBJ 模型计算直接光照，最终效果如图 10-5 所示。

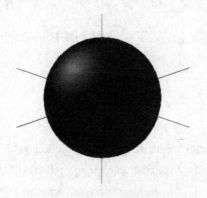

Blinn-Phong 光照模型

图 10-5　Blinn-Phong 光照模型效果图

10.3　明暗过渡计算

10.3.1　Flat 明暗过渡

Flat 明暗过渡是一种简单的能量插值过渡方式，即用一种亮度的颜色作为整个(网格)

面的颜色。本实验以顶点为基本单位，先利用光照模型计算出面顶点的能量(颜色)，然后取这个面顶点能量的平均值作为该面囊括的所有点的能量。对于面内点坐标计算，采用双线性插值的方式实现。

1. 关键代码实现

为了实现对边的插值点计算，此处取 p_1、p_2、p_3 表示三角形网格面的三个顶点；rate 是控制插值密度的变量，当 t 取遍[0,1]时，即实现对边的离散采样。代码如下：

```
for (float t = 0; t <= 1; t += rate)
{
    //边上插值点坐标
    my_3D_point_coord point1((t * p1.x + (1-t) * p2.x), (t * p1.y + (1-t) * p2.y),
                              (t * p1.z + (1-t) * p2.z));
    my_3D_point_coord point2((t * p1.x + (1-t) * p3.x), (t * p1.y + (1-t) * p3.y),
                              (t * p1.z + (1-t) * p3.z));

    //面内部点插值
    for (float u = 0; u <= 1; u += rate)
    {
        //点的位置
        my_3D_point_coord point3((u * point1.x + (1-u) * point2.x),
                (u * point1.y + (1-u) * point2.y), (u * point1.z + (1-u) * point2.z));

        //average_r, average_g, average_b 表示已知的三角形面三个顶点颜色的平均值
        //对三角形面内部所有插值点进行相同颜色赋值
        glBegin(GL_POINTS);
            glColor3f(average_r, average_g, average_b);
            glVertex3f(point3.x, point3.y, point3.z);
        glEnd();
    }
}
```

2. 案例效果

根据上述插值方式，使用 Flat 明暗过渡给 OBJ 模型加上直接光照，最终效果如图 10-6 所示。

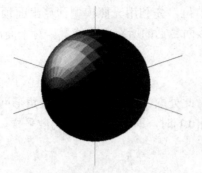

Flat 明暗过渡

图 10-6 Flat 明暗过渡效果图

10.3.2 Gouraud 明暗过渡

Gouraud 明暗过渡，通过对模型网格面上顶点的能量进行双线性插值实现模型表面颜色的明暗过渡，具体过程如下：

(1) 对于网格面顶点，通过读取 OBJ 模型的信息来获得顶点的法向信息。
(2) 利用光照模型，计算该点的光照能量(颜色)。
(3) 利用双线性插值，插值网格面内部点的坐标和能量。

1. 关键代码实现

代码如下：

```
for (float t = 0; t <= 1; t += rate)
{
    //边上能量插值
    float point1Color_r = t * p1Color_r + (1-t) * p2Color_r;
    float point1Color_g = t * p1Color_g + (1-t) * p2Color_g;
    float point1Color_b = t * p1Color_b + (1-t) * p2Color_b;

    float point2Color_r = t * p1Color_r + (1-t) * p3Color_r;
    float point2Color_g = t * p1Color_g + (1-t) * p3Color_g;
    float point2Color_b = t * p1Color_b + (1-t) * p3Color_b;

    //点的位置
    my_3D_point_coord point1((t * p1.x + (1-t) * p2.x), (t * p1.y + (1-t) * p2.y),
                              (t * p1.z + (1-t) * p2.z));
    my_3D_point_coord point2((t * p1.x + (1-t) * p3.x), (t * p1.y + (1-t) * p3.y),
```

(t * p1.z + (1-t) * p3.z));

```
//面内部能量插值
for (float u = 0; u <= 1; u += rate)
{
    //能量插值
    float point3Color_r = u * point1Color_r + (1-u) * point2Color_r;
    float point3Color_g = u * point1Color_g + (1-u) * point2Color_g;
    float point3Color_b = u * point1Color_b + (1-u) * point2Color_b;

    //点的位置
    my_3D_point_coord point3((u * point1.x + (1-u) * point2.x),
        (u * point1.y + (1-u) * point2.y), (u * point1.z + (1-u) * point2.z));

    glBegin(GL_POINTS);                //开始绘制
        glColor3f(point3Color_r, point3Color_g, point3Color_b);
        glVertex3f(point3.x, point3.y, point3.z);
    glEnd();
}
}
```

2. 案例效果

根据上述插值方式，使用 Gouraud 明暗过渡给 OBJ 模型加上直接光照，最终效果如图 10-7 所示。

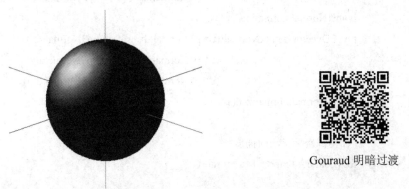

图 10-7 Gouraud 明暗过渡效果图

10.3.3 Phong 明暗过渡

Phong 明暗过渡，通过对模型网格面上点的法向量进行双线性插值实现模型表面颜色的明暗过渡，具体过程如下：

(1) 通过读取 OBJ 模型的信息来获得网格面顶点及其法向信息。
(2) 利用双线性插值，插值网格面的所有边以及内部点的坐标和法向。
(3) 采用光照模型，直接计算上述所有点的光照能量(颜色)。

与 Gouraud 明暗过渡类似，在 Phong 明暗过渡中，同样使用双线性插值来实现，但区别在于，该方法通过插值点的法向量以及直接光照能量计算来实现模型表面的明暗过渡。

1. 关键代码实现

代码如下：

```
for (float t = 0; t <= 1; t += rate)
{
    //边上插值点坐标
    my_3D_point_coord point1((t * p1.x + (1-t) * p2.x), (t * p1.y + (1-t) * p2.y),
                             (t * p1.z + (1-t) * p2.z));
    my_3D_point_coord point2((t * p1.x + (1-t) * p3.x), (t * p1.y + (1-t) * p3.y),
                             (t * p1.z + (1-t) * p3.z));

    //插值相应点的法向量
    my_3Dvector point1Normal((t * p1Normal.dx + (1-t) * p2Normal.dx),
                             (t * p1Normal.dy + (1-t) * p2Normal.dy),
                             (t * p1Normal.dz + (1-t) * p2Normal.dz));
    point1Normal.normalized();
    my_3Dvector point2Normal((t * p1Normal.dx + (1-t) * p3Normal.dx),
                             (t * p1Normal.dy + (1-t) * p3Normal.dy),
                             (t * p1Normal.dz + (1-t) * p3Normal.dz));
    point2Normal.normalized();

    //面内部坐标、法向插值
    for (float u = 0; u <= 1; u += rate)
    {
        //插值的点坐标
```

第 10 章 光 照 计 算

```
    my_3D_point_coord point3((u * point1.x + (1-u) * point2.x),
        (u * point1.y + (1-u) * point2.y), (u * point1.z + (1-u) * point2.z));

    //插值的法向坐标
    my_3Dvector point3Normal((u * point1Normal.dx + (1-u) * point2Normal.dx),
                             (u * point1Normal.dy + (1-u) * point2Normal.dy),
                             (u * point1Normal.dz + (1-u) * point2Normal.dz));
    point3Normal.normalized();

    /*利用光照模型,使用上述点坐标和法向,计算 point3 的能量(颜色):total_r、
    total_g、total_b*/

    glBegin(GL_POINTS);                    //开始绘制
        glColor3f(total_r, total_g, total_b);
        glVertex3f(point3.x, point3.y, point3.z);
    glEnd();
    }
}
```

2. 案例效果

根据上述插值方式,使用 Phong 明暗过渡给 OBJ 模型加上直接光照,最终效果如图 10-8 所示。

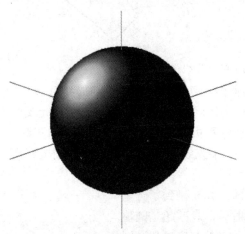

Phong 明暗过渡

图 10-8　Phong 明暗过渡效果图

10.4 全局光照之光线跟踪算法

直接光照模型并不能很好地模拟光的折射、反射及阴影等，也难以反应物体间的相互映射。为解决上述问题，本节实验引入经典的全局光照算法——光线跟踪算法。光线跟踪算法是真实感图形学中的重要算法之一，该算法综合考虑了光的反射、折射、阴影等，具有原理简单、实现方便和能够生成各种逼真的视觉效果等突出的优点。在本节中，仅考虑光的反射和折射。

在如图 10-9 所示的场景中，有一个视点，两个物体 O_1 与 O_2。首先，从视点出发经过投影平面上一点的视线，传播到达球体 O_1，交点为 P_1。利用直接光照模型如式(10.4)可以得到 P_1 点处的光照能量。在 P_1 点，这条光线产生了反射光线 R_1 和折射光线 T_1。

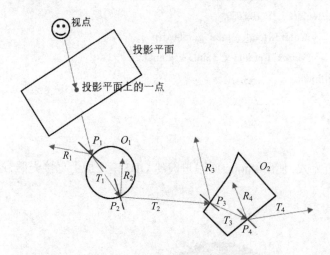

图 10-9 光线跟踪过程示意图

在图 10-9 中，反射光线 R_1 没有再与其他物体相交，则放弃对该条光线的跟踪。折射光线 T_1 在物体 O_1 内部传播，与 O_1 相交于点 P_2。在 P_2 点，这条光线产生了反射光线 R_2 和折射光线 T_2。同理，在反射光线 R_2 和折射光线 T_2 可以继续递归跟踪，并计算交点处的光照能量。考虑到效率上的需要，一条光线(包括折射光线和反射光线)可以有以下几种终止条件：

(1) 该光线未碰到任何物体，即该光线碰到了背景。
(2) 光线在经过多次反射和折射后，衰减超过阈值。
(3) 光线反射或者折射次数，即跟踪深度，超过阈值。

1. 关键代码实现

代码如下：

1. 依据图形显示窗口在横向和纵向的像素点数量对投影面实施点采样。
2. 光线跟踪(循环遍历投影面所有采样点)
   {
   2.1 从视点沿采样点向场景发射射线(入射光线)。
   2.2 计算光线与场景中所有物体的相交情况，找出离采样点最近的交点。
   2.3 利用光照模型计算该交点的直接光照能量(颜色)。
   2.4 若有反射系数，则计算反射光线方向，并以该交点为起点，对反射光线进行光线跟踪，获得反射光返回的光照能量(颜色)。
   2.5 若有折射系数，则计算折射光线方向，并以该交点为起点，对折射光线进行光线跟踪，获得折射光返回的光照能量。
   2.6 返回上述光照能量(颜色)=直接光颜色 + 折射光返回颜色 * 折射系数 + 反射光返回颜色 *反射系数。
   }

2. 案例效果

使用 Blinn-Phong 光照模型对 OBJ 模型进行直接光照计算，最终效果如图 10-10 所示；加上光线跟踪后，生成的最终渲染效果如图 10-11 所示。

图 10-10　Blinn-Phong 光照模型效果图　　图 10-11　光线跟踪效果图　　全局光照之光线跟踪算法

10.4.1　光线跟踪中的求交计算

由于每一条光线都需要与场景中的所有模型进行求交，并找到离投影面最近的一个交点，因此，求交计算是影响光线跟踪效率的关键部分。本实验使用参数方程来表示每条光线，并采用优化的线面求交方法来加速上述过程。

10.4.2　反射光线方向的计算

当光线击中物体表面的一点时，如果物体表面存在反射系数，则需要模拟反射光线，对该光线实施跟踪。根据相关文献可知，反射光线方向 = 入射光向量 −2×交点法向×(入射光向量与交点法向的点乘)。

10.4.3 折射光线方向的计算

当光线击中物体表面的一点时，如果物体表面存在折射系数，则需要模拟折射光线，对该光线实施跟踪。根据相关文献可知，折射光线方向 Refr_Dir 由如下公式计算获得：

$$\text{Refr\_Dir} = \frac{\eta_i}{\eta_t}\left(L - (L \cdot N) \times N\right) - \sqrt{1 - \left(\frac{\eta_i}{\eta_t}\right)^2\left(1 - (L \cdot n)^2\right)} \times N \tag{10.6}$$

其中：L 为入射光向量；N 为交点法向；η_i 和 η_t 分别表示光线所离开的物体反射率和所进入的物体反射率。

对应代码实现如下：

```
/*****************************************
* 计算折射光线方向的函数
* inpuray_dir：入射光向量；nhit：交点法向；refracted_dir：输出的折射光线方向；
* ni_over_nt：所离开物体与被进入物体之间的反射率比值，用斯奈尔定律计算
******************************************/
bool get_refract_dir_my(const my_3Dvector& inpuray_dir,  my_3Dvector& nhit,
                   float ni_over_nt, my_3Dvector& refracted_dir)
{
    my_3Dvector uv(inpuray_dir.dx,inpuray_dir.dy,inpuray_dir.dz);
    uv.normalized();
    float dt = uv.dot(nhit);
    float discriminant = 1.0 - ni_over_nt * ni_over_nt * (1-dt * dt);
    if (discriminant > 0)
    {
        refracted_dir = (uv - nhit * dt) * ni_over_nt - nhit * sqrt(discriminant);
        return true;
    }
    else
        return false;
}
```

10.5　课外拓展性实验

任务 1：在△ABC 中，A、B、C 三点的坐标分别为(0,0,0)、(5,0,0)、(0,5,0)，三点的能量值分别为 100、150、200。现有一点 $P(2,2,0)$，请利用双线性插值的方法，计算 P 点的能量值。

任务 2：结合 10.4 节所学，加载球体模型。
(1) 利用不同的光照模型实现直接光照计算，并对比不同光照模型的差别。
(2) 通过对模型进行双线性点插值，实现模型的明暗过渡，并对比不同明暗过渡之间的差别。

任务 3：加载复杂场景模型，开展光线跟踪。

第 11 章 纹 理 映 射

11.1 实验内容简述和实验目标

基本实验内容包括：纹理映射(创建纹理、设置纹理参数、设定映射方式、映射纹理坐标与几何坐标、设置纹理的重复度)和 Shadow Mapping 阴影生成。同时，配备了一个课外拓展性实验。

完成本实验后，读者能够：
(1) 熟记和描述全局光照的基本原理和光线跟踪算法(布鲁姆知识模型：记忆、理解)；
(2) 熟记和描述阴影的产生原理(布鲁姆知识模型：记忆、理解)；
(3) 结合 OpenGL 能在静态场景下基于光线跟踪进行绘制(布鲁姆知识模型：应用)。

11.2 纹 理 映 射

纹理映射是真实感图形制作中非常重要的一个环节，其过程较复杂，包括：① 创建纹理；② 设置纹理参数；③ 设定映射方式；④ 映射纹理坐标和几何坐标；⑤ 设置纹理的重复度。

11.2.1 创建纹理

本小节的纹理创建开始于本地图像的加载，使用 LoadImage 函数，其原型为
  HANDLE LoadImage(NINSTANCE hinst, LPCTSTR lpszName, UINT uType,
    int cxDesired, int cyDesired, UINT fuLoad);
其中：参数 hinst 用于指示图像资源的来源，若属于当前程序以外的资源，则其值取 NULL，否则其值取 Windows 标准函数 AfxGetInstanceHandle 的返回值；参数 lpszName 是图像在本地存放的完整路径，包括文件名；参数 uType 用于指定图像类型，如 IMAGE_BITMAP；参数 cxDesired、cyDesired 分别指定图像实际的长和宽；参数 fuLoad 表示额外的属性配置，

一般为 LR_DEFAULTCOLOR，表示缺省属性。更多其他格式图像的加载，读者可参考相关文献。

加载本地图像后，需要将其创建成在 OpenGL 中能够进行映射的纹理，即将图像设置为当前 OpenGL 渲染管线中的纹理。针对二维纹理，这里使用 glTexImage2D 函数，其原型为

void glTexImage2D(GLenum target, GLint level, GLint components, GLsizei width, Glsizei height, GLint border, GLenum format, GLenum type, const GLvoid *pixels);

其中：参数 target 用于指定纹理类别，它可以是 GL_TEXTURE_1D、GL_TEXTURE_2D 和 GL_TEXTURE_3D，分别对应一维、二维和三维纹理，此处设置为 GL_TEXTURE_2D；参数 level 表示多级分辨率的纹理图像的级数，若只有一种分辨率，则 level 设置为 0；参数 components 表示用图像中的哪些分量进行调整和混合，取值为 1 到 4 的整数，1 表示选择了 R 分量，2 表示选择了 R 和 A 两个分量，3 表示选择了 R、G、B 三个分量，4 表示选择了 R、G、B、A 四个分量；参数 width 和 height 分别给出了图像的长度和宽度单位为像素；参数 border 指纹理边界的长度，一般取两个像素；参数 format 和 type 分别描述了纹理映射的格式和数据类型；参数 pixels 为通过 LoadImage 加载进来的图像数据。

11.2.2 设置纹理参数

纹理映射时，由于图像和待映射多边形区域在尺寸和形状上存在差异，因此会造成映射到多边形的图像走样。为改善上述效果，可用 glTexParmeter{if} 函数来调整映射模式，该函数原型为

void glTexParameter{if} (GLenum target, GLenum pname, GLint param);

其中：参数 target 用于指定纹理类型，通常设置为 GL_TEXTURE_2D；参数 pname 用于指定纹理映射的滤波(插值)类型或重复模式(当 pname 为滤波类型时，通常有 GL_TEXTURE_MAG_FILTER 和 GL_TEXTURE_MIN_FILTER 两种类型，分别对应纹理被放大时的滤波和纹理被缩小时的滤波；当 pname 为重复模式时，表示在重复映射的情况下，纹理在 s、t 方向上的重复模式，取值有 GL_TEXTURE_WRAP_S 和 GL_TEXTURE_WRAP_T)；参数 param 指 pname 的取值(当 pname 为滤波类型时，通常取 GL_LINEAR(即线性插值或加权平均插值，插值速度较慢但无锯齿，结果往往比较模糊或朦胧)或 GL_NEAREST(采用就近原则选取像素，速度快，但走样比较明显)；而当 pname 为重复模式时，通常取 GL_CLAMP 或 GL_REPEAT)。

11.2.3 设定映射方式

OpenGL 提供了多种纹理映射模式，既可以将图像颜色直接映射到多边形上，又可以

用纹理中的值来调整多边形表面的原有颜色,即将图像中的颜色与多边形原来的颜色进行混合。用于设置上述映射模式的函数如下:

void glTexEnv{if}(GLenum target, GLenum pname, TYPE param);

其中,参数 target 取值固定为 GL_TEXTURE_ENV。若参数 pname 为 GL_TEXTURE_ENV_MODE,则参数 param 可以为 GL_DECAL、GL_MODULATE 或 GL_BLEND;若参数 pname 为 GL_TEXTURE_ENV_COLOR,则参数 param 是包含四个浮点数(分别对应 R、G、B 和 A 的取值)的数组。

11.2.4 映射纹理坐标与几何坐标

在绘制带纹理的三维图形时,需要给三维图形(以网格面为单位)的每个顶点确定纹理坐标。顶点坐标决定了顶点在屏幕上的像素坐标,纹理坐标决定了纹理(图像)中的哪一个纹素(颜色)对应到该像素坐标上。同时,三维图形每个网格面内部点(最后也对应到屏幕上的像素)的纹理坐标可采用第 10 章光照明暗过渡中的插值方法来确定。

为了便于开展插值和纹理映射,无论加载的原始图像长宽如何,在进入纹理映射前,通常都被转化为一张在 s 方向(横向,对应 x 轴)和 t 方向(纵向,对应 y 轴)上都被归一化的纹理(图像)。特别地,有些参考资料中的纹理坐标系用 u 和 v 表示,u 对应 s,v 对应 t。在 OpenGL 中指定几何点对应纹理坐标的常用函数如下:

void glTexCoord2{sifd} (TYPE coords);

其中,参数 coords 为纹理坐标系下的坐标,即 s 和 t(或 u 和 v)的取值。之后,该指定的纹理坐标会与该函数后面紧跟的绘制顶点 glVertex{ifd}建立对应关系,即完成了 OpenGL 环境下,纹理坐标与几何坐标的对应。例如,以下两行代码表示纹理坐标为(0.0, 0.0)的颜色与三维坐标点(−50.0, −50.0, 0.0)之间建立映射关系,即当三维点被绘制到屏幕时,所呈现的颜色为(0.0, 0.0)纹理像素值:

glTexCoord2f(0.0, 0.0);

glVertex3f(-50.0, -50.0, 0.0);

特别地,对于复杂的模型表面,通常 OpenGL 还提供了纹理坐标和几何坐标的对应关系自动设置函数 glTexGen(在本书中不做讨论)。

1. **案例关键代码实现**

结合上述纹理映射的四个步骤,在 OpenGL 平台下实现纹理映射的关键代码如下:

glGenTextures(1, &textName);   //创建纹理 textName 的调用句柄,用于调用
                                      //加载图片
HBITMAP hBMP = (HBITMAP)LoadImage(NULL, 本地图片, IMAGE_BITMAP, 0, 0,

```
                              LR_CREATEDIBSECTION | LR_LOADFROMFILE);
BITMAP BMP;
GetObject(hBMP, sizeof(BMP), &BMP);
glPixelStorei(GL_UNPACK_ALIGNMENT, 4);


//进行纹理绑定和参数设置
glBindTexture(GL_TEXTURE_2D, textName);
glTexParameteri(GL_TEXTURE_2D, GL_TEXTURE_MAG_FILTER, GL_LINEAR);
glTexParameteri(GL_TEXTURE_2D, GL_TEXTURE_MIN_FILTER, GL_LINEAR);
glTexImage2D(GL_TEXTURE_2D, 0, 3, BMP.bmWidth, BMP.bmHeight, 0,
             GL_BGR_EXT, GL_UNSIGNED_BYTE, BMP.bmBits);


//设置纹理映射方式(混合模式)
glTexEnvf(GL_TEXTURE_ENV, GL_TEXTURE_ENV_MODE, GL_MODULATE);


//实施纹理坐标和几何坐标映射
glBindTexture(GL_TEXTURE_2D, textName);
glBegin(GL_TRIANGLES);                //绘制三角面片
    glTexCoord2f(p1TextCoord.u, p1TextCoord.v);
    glVertex3f(p1.x, p1.y, p1.z);


    glTexCoord2f(p2TextCoord.u, p2TextCoord.v);
    glVertex3f(p2.x, p2.y, p2.z);


    glTexCoord2f(p3TextCoord.u, p3TextCoord.v);
    glVertex3f(p3.x, p3.y, p3.z);
glEnd();
```

2. **案例效果**

如图 11-1 所示为未贴纹理的矩形网格面绘制效果图,图 11-2 所示为对上述网格面实施纹理映射后所得到的网格面绘制效果图。

映射纹理坐标与几何坐标

图 11-1　直接网格面绘制效果图　　图 11-2　纹理映射绘制效果图

11.2.5　设置纹理的重复度

对于给定的网格面，其纹理的重复度，除了通过 glTexParameter* 函数进行调整之外，在进行纹理坐标和模型几何坐标对应时也可以改变。例如，在纹理坐标的 s 和 t 方向上都从标准的[0,1]扩大到[0,2]（即 glTexCoord* 中的参数值），并保持待映射网格面的尺寸不变（即 glVertex3f 中的坐标值）。此时，在 s 和 t 方向上，纹理的密度均提升了两倍，如图 11-3 和图 11-4 所示。

 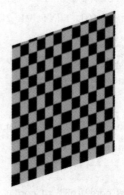

图 11-3　初始纹理映射效果图　　图 11-4　改变纹理映射重复度的效果图

1. 关键比较代码

代码如下：

```
//代码段 1：初始纹理坐标和几何坐标映射代码，对应效果见图 11-3
glBegin(GL_QUADS);                //绘制矩形
    glTexCoord2f(0.0f, 0.0f); glVertex3f(200, 0, 0);
    glTexCoord2f(1.0f, 0.0f); glVertex3f(200, 300, 0);
```

```
        glTexCoord2f(1.0f, 1.0f); glVertex3f(200, 300, 250);
        glTexCoord2f(0.0f, 1.0f); glVertex3f(200, 0, 250);
    glEnd();

//代码段2：增加重复度的纹理坐标和几何坐标映射代码，对应效果见图11-4
    glBegin(GL_QUADS);              //绘制矩形
        glTexCoord2f(0.0f, 0.0f); glVertex3f(200, 0, 0);
        glTexCoord2f(2.0f, 0.0f); glVertex3f(200, 300, 0);
        glTexCoord2f(2.0f, 2.0f); glVertex3f(200, 300, 250);
        glTexCoord2f(0.0f, 2.0f); glVertex3f(200, 0, 250);
    glEnd();
```

设置纹理的重复度

2. 案例效果

在纹理坐标绑定过程中，将 s 和 t 方向均设为标准的[0,1]，此时纹理映射效果如图11-3所示。当将 s 和 t 方向的坐标提升到[0,2]时，纹理映射效果如图11-4所示。

11.3 Shadow Mapping 阴影生成

为了使读者了解阴影的生成过程，本节开展基础的 Shadow Mapping 阴影生成实验。假定视点坐标系架设在 z 轴正方向上，视线方向朝向 z 轴负方向，投影面与 xoy 坐标轴平面平行，且处于 z 轴正方向上，点光源放置于 y 轴正方向上。Shadow Mapping 阴影生成的算法步骤如下：

(1) 根据图形显示窗口中像素点在横向和纵向的分布数量，对投影面进行均匀点采样。

(2) 将视点坐标系进行旋转，使得视点与点光源重合，再从光源沿着投影面(已随视点坐标系同步旋转)的每个采样点向场景发射射线，与场景求交，并取得距离光源最近的点存储到深度缓存矩阵中(矩阵中的每个单元存放一个三维点坐标和相应的阴影值)，同时在该投影面设置一个与其位置重合的虚拟的深度缓存面。

(3) 将视点坐标系反向旋转到初始位置，利用光线跟踪算法(仅跟踪一次)对场景进行阴影计算。具体的步骤如下：

① 从视点沿着投影面(已随视点坐标系同步旋转回归原位)的每个采样点向场景发射射线(入射光线)。

② 计算上述每条光线与场景中所有物体的相交情况，找出离采样点最近的交点。

③ 将该交点与光源相连，计算其与深度缓存面的交点。

④ 在深度缓存矩阵中寻找与该交点最近的矩阵单元，将其阴影值或其周围若干矩阵单元中的阴影平均值作为该交点的阴影值。

1. 关键数据结构

(1) 对原有的三维点数据结构进行扩展(代码如下所示)，并将其用于本节的阴影生成实验。

```
struct my_3D_point_coord_Ex
{
    float x;
    float y;
    float z;
    my_3D_point_coord_Ex add(float xi, float yi, float zi);    //顶点坐标加法运算
};
```

(2) 对原有的三维向量数据结构进行扩展(代码如下所示)，并将其用于本节的阴影生成实验。

```
class my_3Dvector_Ex
{
    public:
    float dx;
    float dy;
    float dz;
    float len;        //向量的长度
    public:
    my_3Dvector_Ex(my_3D_point_coord_Ex start, my_3D_point_coord_Ex end);
    my_3Dvector_Ex cross(const my_3Dvector_Ex& input_vector);    //向量叉乘
    float dot(const my_3Dvector_Ex& input_vector);                //向量点乘
    my_3Dvector_Ex operator +(const my_3Dvector_Ex& input_vector); //向量相加
    my_3Dvector_Ex operator -(const my_3Dvector_Ex& input_vector); //向量相减
    my_3Dvector_Ex operator * (const float& cons);                //向量乘以常数
    float length();                                               //获得向量本身的长度
    void normalized();                                            //对向量归一化
};
```

(3) 深度缓存矩阵单元定义如下：

```
struct my_map
{
```

```
    float shadow = 0;              //初始化点为黑色，即处于阴影中，取值为 0~1 的浮点数
    float t = 999999;              //用作记录采样点和模型上的点的距离，初始化为最大
    my_3D_point_coord_Ex models_crossed_point;   //透过采样点的光线与模型相交的点
};
```

(4) 三维空间中射线的数据结构定义如下：

```
class my_3D_line
{
public:
    my_3D_point_coord_Ex my_orig;              //起点
    my_3Dvector_Ex my_dir;                     //方向向量
public：
    my_3D_line();
    my_3D_line(const my_3D_point_coord& orig, const my_3Dvector& direction);
    my_3D_line(float origx, float origy, float origz, float raydirx, float raydiry,
                float raydirz);
};
```

(5) 三维模型每个三角形网格面的数据结构定义如下：

```
struct my_3D_triangle
{
    my_3D_point_coord_Ex first_point;          //三角形上的第一个点
    my_3D_point_coord_Ex second_point;         //三角形上的第二个点
    my_3D_point_coord_Ex third_point;          //三角形上的第三个点
};
```

2. 关键函数代码实现

代码如下：

```
/***********************************
 * 对投影面进行离散点采样
 * 本函数要求视点坐落于世界坐标系的某条坐标轴上，且投影平面与上述坐标轴垂直
 * 上述坐标轴垂直于上述坐标平面
 * left_x、right_x、bottom_y、up_y：投影面(视口)矩形四个方向的边界值；
 * nearplane_distance：近平面距离；
 * eye_z：视点在 z 轴上的值；
 * render_vertices：采样点集合；
```

* width、height：分别记录横向和纵向采样点的个数

************************************/

```cpp
void samplepoint_sonprojectionplan(float left_x, float right_x, float bottom_y, float up_y,
            float nearplane_distance, float eye_z,
            map<my_3D_point_coord_Ex*, my_draw_color*>& render_vertices,
            unsigned& width, unsigned& height)
{
    //对投影区域进行采样密度设置
    float x_delt = (right_x - left_x) / (ceil(right_x) - floor(left_x));
    float y_delt = (up_y - bottom_y) / (ceil(up_y) - floor(bottom_y));
    float z_val = eye_z - nearplane_distance - 1;

    width = 0; height = 0;
    bool counted = false;
    for (float x_iter = left_x; x_iter <= right_x; x_iter += x_delt)
    {
        width++;
        for (float y_iter = bottom_y; y_iter <= up_y; y_iter += y_delt)
        {
            my_3D_point_coord_Ex* tempPoint_ptr = new
                my_3D_point_coord_Ex(x_iter, y_iter, z_val);
            my_draw_color* tempColor_ptr = new my_draw_color{ 0,0,0 };
            render_vertices.insert(pair<my_3D_point_coord_Ex*,
                my_draw_color*>(tempPoint_ptr, tempColor_ptr));

            if (counted == false) height++;
        }
        counted = true;
    }
}
```

/************************************

* 构建深度缓存矩阵

* 本函数要求视点坐落于世界坐标系的某条坐标轴上,且投影平面与上述坐标轴垂直
* 上述坐标轴垂直于上述坐标平面
* left_x、right_x、bottom_y、up_y:投影面(视口)矩形四个方向的边界值;
* light_position:光源位置;
* all_models:场景中的所有三维图形
**************************************/
```
void build_shadow_map (float left_x, float right_x, float bottom_y, float up_y,
                       my_3D_point_coord_Ex& light_position,
                       const vector<my_triangle_3DModel>& all_models)
{
    float x_delt , y_delt;        //直接指定采样面点之间的间隔或密度
    int max_i, max_j             //采样横向和纵向的数量

    for (int i = 0; i < max_i; i++)
    {
        for (int j = 0; j < max_j; j++)
        {
            my_3D_point_coord_Ex samplepoint(left_x + i * x_delt, 1000,
                                              bottom_y + j * y_delt);     //采样点
            my_3D_point_coord_Ex lightposition(light_position.x, light_position.y,
                                                light_position.z);        //光源坐标
            my_3Dvector_Ex lightvector(lightposition, samplepoint);       //射线向量
            lightvector.normalized();
            my_3D_line light_line(lightposition, lightvector);            //生成射线

            //遍历 all_models 中的每个三维图形并与 light_line 求交,确定离光源最近的交点;
            //将该交点存在深度缓存矩阵 my_map_point[i][j]中
        }
    }
}
```

/***********************************

```
 * 计算阴影
 * left_x、right_x、bottom_y、up_y：投影面(视口)矩形四个方向的边界值；
 * rayorig、raydir：分别为光线发出的位置和朝向；
 * all_models：场景中的所有三维图形；
 * eye_position：视点位置；
 * light_position：光源位置；
 * my_map_point：深度缓存矩阵(二维数组)，为全局变量，类型为 my_map
 * 返回当前点的阴影值
**************************************/
float one_ray_trace_shadow(float left_x, float right_x, float bottom_y, float up_y,
                my_3D_point_coord_Ex rayorig, my_3Dvector_Ex raydir,
                const vector<my_triangle_3DModel>& all_models,
                const my_3D_point_coord_Ex& eye_position,
                const my_3D_point_coord_Ex& light_position)
{
    //采用三角网格遍历计算模型与光线的交点
    float nearest_t = INFINITY;       //射线参数方程中的 t
    my_3Dvector_Ex nearestTrangleNormal;
    const my_triangle_3DModel* nearestModel = NULL;
    my_3D_line curRay(rayorig.x, rayorig.y, rayorig.z, raydir.dx, raydir.dy, raydir.dz);
    for (unsigned model_index = 0; model_index < all_models.size(); model_index++)
    {
        for (unsigned int tri_index=0; tri_index <
        all_models[model_index].faceSets.size(); tri_index++)
        {
            //从面实例中取出三个顶点
            int firstPointIndex =
                    all_models[model_index].faceSets[tri_index].first_point_index;
            int secondPointIndex =
                    all_models[model_index].faceSets[tri_index].second_point_index;
            int thirdPointIndex =
                    all_models[model_index].faceSets[tri_index].third_point_index;
```

```
my_3D_point_coord_Ex p1 =
    all_models[model_index].pointSets[firstPointIndex];      //第一个顶点
my_3D_point_coord_Ex p2 =
    all_models[model_index].pointSets[secondPointIndex];     //第二个顶点
my_3D_point_coord_Ex p3 =
    all_models[model_index].pointSets[thirdPointIndex];      //第三个顶点

my_3D_triangle curTriangle = { p1, p2, p3};
IntersectionBetweenLineAndTriangle newIntTest(curRay, curTriangle);

//不仅有交点,还要求不能是出发点附近距离 0.002 范围内的点
if (newIntTest.Find() && newIntTest.GetLineParameter() > 0.002f &&
    (newIntTest.GetLineParameter() < nearest_t))
{
    nearestModel = &all_models[model_index];
    nearestTrangleNormal = newIntTest.GetHitPointNormal();
    nearest_t = newIntTest.GetLineParameter();
}
        }
    }
}

//若当前光线与模型无交,则表示光线击中了空间背景,无阴影,返回白色
if (!nearestModel)
    return 1.0;

my_3Dvector_Ex added_valVec = raydir * nearest_t;
my_3D_point_coord_Ex phit = rayorig.add(added_valVec.dx, added_valVec.dy, added_valVec.dz);
                                        //获得交点
my_3Dvector nhit = nearestTrangleNormal;         //获得交点处的法向
nhit.normalized();

int i, j;
float shadow_color = 0;
```

```
my_3D_point_coord_Ex crosspoint;            //与深度缓存面的交点
my_3Dvector_Ex light_to_point(light_position, phit);

//求得点与深度缓存面的交点
crosspoint = CalPlaneLineIntersectPoint(my_3Dvector_Ex(0, -1, 0),
my_3D_point_coord_Ex(-left_x, 1000, -bottom_y), light_to_point, phit);

//判断交点在数组上的大致位置
if (crosspoint.x<left_x || crosspoint.x>right_x || crosspoint.z<bottom_y || crosspoint.z>up_y)
                    //在缓存面范围外,即光源无法覆盖的区域,返回黑色
{
    return 0;
}
else                //与深度缓存面有实交点
{
    float ii = (crosspoint.x - left_x) / deltx;      //横向采样间隔
    float jj = (crosspoint.z - bottom_y) / delty;    //纵向采样间隔
    int ii_up = ceil(ii);                            //i 的上界
    int ii_down = floor(ii);                         //i 的下界
    int jj_up = ceil(jj);
    int jj_down = floor(jj);
    //刚好在 my_map_point 中找到相应的点
    if (ii_down == ii_up && jj_down == jj_up)
    {
        i = ii_up, j = jj;
        if (distance(phit, my_map_point[i][j].models_crossed_point) <= 1)
        {
            shadow_color = my_map_point[i][j].shadow;
        }
    }
    else         //找到周围阈值范围内的点(最多 4 个),计算平均阴影值
    {
        int count = 0;
```

```
                if (distance(phit, my_map_point[ii_up][jj_up].models_crossed_point) < 5)
                    {shadow_color += my_map_point[ii_up][jj_up].shadow; count++;}

                if (distance(phit, my_map_point[ii_up][jj_down].models_crossed_point) < 5)
                    {shadow_color += my_map_point[ii_up][jj_down].shadow; count++;}

                if(distance(phit,my_map_point[ii_down][jj_down].models_crossed_point)<5)
                    {shadow_color += my_map_point[ii_down][jj_down].shadow; count++;}

                if (distance(phit, my_map_point[ii_down][jj_up].models_crossed_point) < 5)
                    {shadow_color += my_map_point[ii_down][jj_up].shadow; count++;}

                shadow_color /= count;
            }
        }
        return shadow_color;
    }
```

3. 案例效果

如图 11-5 所示为仅有立方体和平面的简单场景，图 11-6 所示为对简单场景进行 Shadow Mapping 后产生的阴影效果图。

图 11-5　无阴影的简单场景　　　　图 11-6　用 Shadow Mapping 产生的阴影效果

Shadow Mapping
阴影生成

11.4 课外拓展性实验

加载 OBJ 模型,采用光线跟踪进行颜色纹理和灯光混合(不使用 OpenGL 自带的纹理混合)。如图 11-7 所示为 OpenGL 自动纹理(颜色纹理)映射效果示意图,图 11-8 所示为直接基于光线跟踪算法进行纹理插值映射得到的效果图。

图 11-7　OpenGL 颜色纹理映射效果图　　　图 11-8　光照和纹理混合的效果图

课外拓展性实验

参 考 文 献

[1] WRIGHT R S，HAEMEL J N，SELLERS G，et al.OpenGL 超级宝典[M]. 5 版. 付飞，李艳辉，译. 北京：人民邮电出版社，2020.

[2] 孙家广，胡事民. 计算机图形学基础教程[M]. 2 版. 北京：清华大学出版社，2009.

[3] ELIAS R. A Problem-solving Approach for Computer Graphics[M]. New York：Springer International Publishing，2014.

[4] 倪明田，吴良芝. 计算机图形学[M]. 北京：北京大学出版社，1999.

[5] OpenGL 基础图形编程[Z/OL]. (2012-08-06)[2021-05-14]. https://blog.csdn.net/iduosi/article/details/7835624.

[6] 赵明. 计算机图形学[Z/OL]. (2021-03-22)[2021-05-14]. https://www.icourse163.org/course/CAU-45006.

[7] Weiler-Atherton 算法[Z/OL]. (2014-07-13)[2021-05-14]. https://blog.csdn.net/yangxi_pekin/article/details/37738219?ops_request_misc=%257B%2522request%255Fid%2522%253A%252216137490371678029902214 0%2522%252C%2522scm%2522%253A%252220140713.130102334.pc%255Fall.%2522%257D&request_id=161374903716780299022140&biz_id=0&utm_medium=distribute.pc_search_result.none-task-blog-2~all~first_rank_v2~rank_v29-3-37738219.first_rank_v2_pc_rank_v29_10&utm_term=Weiler_Athenton%E7%AE%97%E6%B3%95.

[8] 扫描线填充算法(有序边表法)[Z/OL]. (2012-03-19)[2021-05-14]. https://blog.csdn.net/orbit/article/details/7368996?ops_request_misc=&request_id=&biz_id=102&utm_term=%E6%89%AB%E6%8F%8F%E7%BA%BF%E5%A1%AB%E5%85%85%E7%AE%97%E6%B3%95&utm_medium=distribute.pc_search_result.none-task-blog-2~all~sobaiduweb~default-6-7368996.first_rank_v2_pc_rank_v29.

[9] 3D 中的 OBJ 文件格式详解[Z/OL]. (2015-11-29)[2021-05-14]. https://blog.csdn.net/u013467442/article/details/50097821.

[10] OpenGL 导入 OBJ 模型[Z/OL]. (2018-08-03)[2021-05-14]. https://www.cnblogs.com/feifanrensheng/p/9416717.html.

[11] Introduction to Ray Tracing: a Simple Method for Creating 3D Images[Z/OL]. (2014-10-30)[2021-05-14]. https://www.scratchapixel.com/lessons/3d-basic-rendering/introduction-to-ray-tracing.

[12] 光线追踪(ray tracing)介绍与细节推导[Z/OL]. (2019-05-03)[2021-05-14]. https://blog.csdn.net/qq_16013649/article/details/89764346?utm_medium=distribute.pc_relevant_t0.none-task-blog-BlogCommendFromMachineLearnPai2-1.control&depth_1-utm_source=distribute.pc_relevant_t0.none-task-blog-BlogCommendFromMachineLearnPai2-1.control.

[13] LoadImage 用法[Z/OL]. (2015-05-15)[2021-05-14]. https://blog.csdn.net/hisinwang/article/details/45752089?ops_request_misc=%257B%2522request%255Fid%2522%253A%2522161233642116780265413040%2522%252C%2522scm%2522%253A%252220140713.130102334..%2522%257D&request_id=161233642116780265413040&biz_id=0&utm_medium=distribute.pc_search_result.none-task-blog-2~all~sobaiduend~default-4-45752089.first_rank_v2_pc_rank_v29_10&utm_term=LoadImage&spm=1018.2226.3001.4187.

[14] 射线与三角面片求交[CP/OL]. [2021-05-14]. https://www.geometrictools.com/GTE/Mathematics/IntrRay3Triangle3.h.

[15] OpenGL 纹理映射总结[Z/OL]. (2011-07-17)[2021-05-14]. https://www.cnblogs.com/sunliming/archive/2011/07/17/2108917.html.

[16] OpenGL 纹理入门[Z/OL]. (2012-07-19)[2021-05-14]. https://blog.csdn.net/hippig/article/details/7764990?utm_medium=distribute.pc_relevant.none-task-blog-baidujs_utm_term-11&spm=1001.2101.3001.4242.

[17] Shadow Mapping 的原理与实践[Z/OL]. (2015-12-03)[2021-05-14]. https://www.cnblogs.com/mazhenyu/p/5015345.html.